HELICOPTERS IN THE HIGH COUNTRY

40 YEARS OF MOUNTAIN FLYING

HELICOPTERS IN THE HIGH COUNTRY

40 YEARS OF MOUNTAIN FLYING

Peter Corley-Smith and David N. Parker

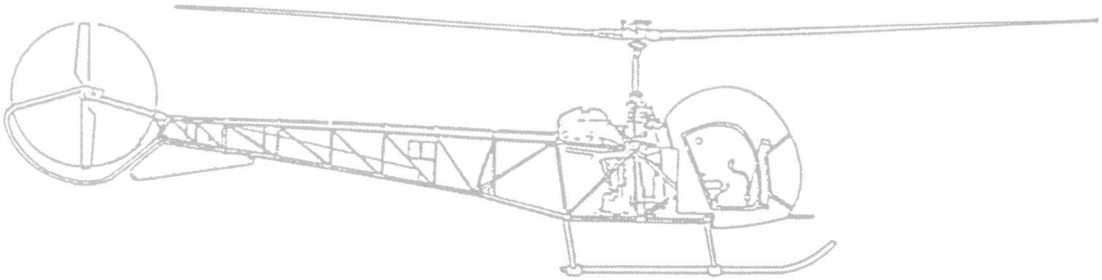

VICTORIA • BRITISH COLUMBIA • CANADA

Sono Nis Press

Excerpts from *Helicopters: The British Columbia Story* (1985) by Peter Corley-Smith and David N. Parker reprinted by permission of the Royal British Columbia Museum.

Canadian Cataloguing in Publication Data

Corley-Smith, Peter, 1923-
Helicopters in the high country

Includes bibliographical references and index.
ISBN 1-55039-061-9

1. Helicopter transportation—British Columbia—History. 2. Helicopters—History. I. Parker, David N. II. Title.
HE9793.C22B7 1995 387.7'3352'09711 C95-910533-6

First printing, 1995
Second printing, 1996

Publication of this book has been financially assisted by the Canada Council Block Grant Program.

Published by
SONO NIS PRESS
1725 Blanshard Street
Victoria, British Columbia V8W 2J8

Printed and bound in Canada by
MORRISS PRINTING COMPANY LTD.
Victoria, British Columbia

CONTENTS

Introduction	7
Acknowledgements	9
Beginnings: Okanagan Air Services Ltd.	11
Competition: Vancouver Island Helicopters Ltd.	29
A Growing Business	47
Ambulance Service: A New Dimension	69
Helicopter Logging: Another New Dimension	77
Epilogue	85
Interviews	87
Bibliography	89
Index	91

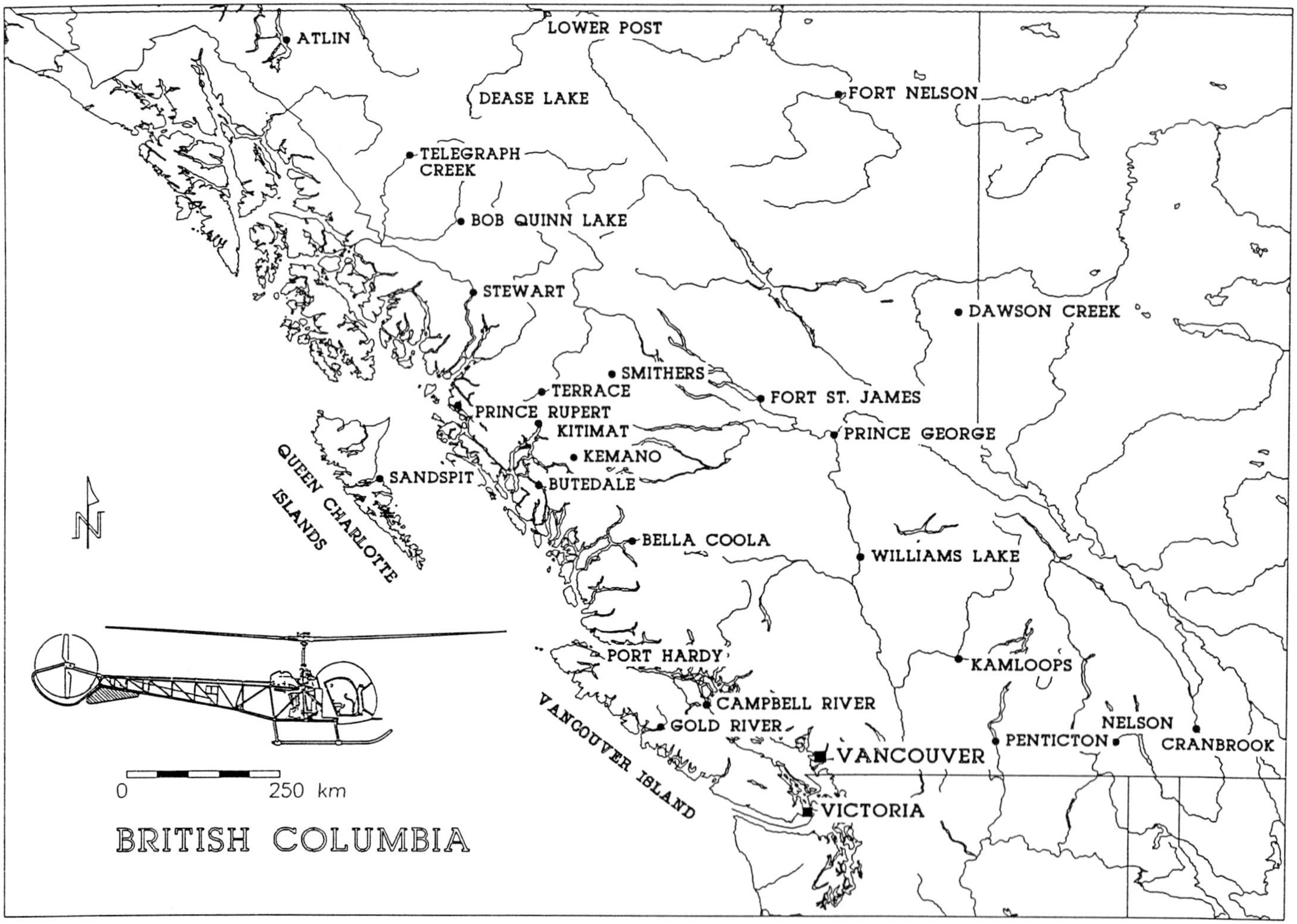
ATLIN
LOWER POST
DEASE LAKE
FORT NELSON
TELEGRAPH CREEK
BOB QUINN LAKE
STEWART
DAWSON CREEK
SMITHERS
TERRACE
FORT ST. JAMES
PRINCE RUPERT
KITIMAT
PRINCE GEORGE
KEMANO
QUEEN CHARLOTTE ISLANDS
SANDSPIT
BUTEDALE
N
BELLA COOLA
WILLIAMS LAKE
PORT HARDY
KAMLOOPS
CAMPBELL RIVER
VANCOUVER ISLAND
GOLD RIVER
NELSON
PENTICTON
CRANBROOK
VANCOUVER
0
250 km
VICTORIA
BRITISH COLUMBIA

INTRODUCTION

The concept of a helicopter can be traced back to Leonardo da Vinci in 1505, who used principles dating all the way back to the design for a "helix" proposed by Archimedes. However, another 400 years were to pass before the first practical results of these theories made their appearance. This happened, more or less at the same time, in Europe and the United States in the late 1930s. Another decade was to pass before the first commercially viable helicopters reached Canada.

In North America, the two most successful designers were Igor Sikorsky and Arthur Young. Sikorsky had tried to develop a helicopter in Russia as early as 1909, but failed to solve the extremely complex aerodynamic and engineering problems involved. Then he emigrated to the United States where he became a very succesful designer of fixed-wing aircraft – particularly his series of flying boats. By 1939, with the backing of United Aircraft Corporation, the parent company of Sikorsky Aircraft, he was able to develop his ideas and produce the first of a remarkably successful line of helicopters: the VS300. Sikorsky did all the developmental test flying himself, and the quaint photograph of him, perched at the front of his machine in his homburg hat, has been published innumerable times.

The other successful designer, Arthur Young, experimented with remote-control models and then took his ideas to Lawrence Bell of the well-established Bell Aircraft Corporation. With Larry Bell's support, Young developed the Bell 30, followed by the successful Bell 47 series. Young, at first, and in spite of his protests, was not allowed to fly his creation: this had to be carried out by the corporations' test pilots – pilots accustomed to high performance pursuit, or fighter aircraft. There is an interesting video of the first of these test-flights in a tethered Model 30. It was in the winter, and a good deal of snow had been cleared off the pad. The pilot lifted off too briskly to about six feet (two m), over-reacted and hit the ground with a thump. Wearing no harness, he was bounced up into the rotor system, which promptly flung him into the snowbank surrounding the pad. He suffered a broken wrist and some bruising but no serious injuries. From then on, Young was allowed to do his own test-flying.

By 1947, the Model 30 had been developed into the 47B. There were two versions: the B, with a closed cabin, and the B3 with an open cockpit. In that same year, the Bell company, whose main plant was in Buffalo, New York, set up a subsidiary, Central Aircraft, to demonstrate its helicopter for agricultural spraying. This company was based in Yakima, Washington, the heart of orchard country – and this is where this story really begins.

The people who took over development of this brand new technology for commercial use needed an unusual spectrum of skills and a great deal of fortitude. To begin with, they had to take serious financial risks; secondly, they needed a generous measure of physical courage, particularly in the mountainous terrain of British Columbia; thirdly, they had to be innovative and pertinacious to persuade a normally reactionary group of businessmen that what they had to offer was not a toy, but would bring a whole new dimension to mineral exploration, surveying and the forest industry (and in the long run, to innumerable other applications).

This is the story of how a few committed people – pioneers such as Alf Stringer, Carl Agar, Barney Bent and Ted and Lynn Henson – set out to open up a whole new industry. No purple prose is needed to illustrate that they were remarkably successful.

ACKNOWLEDGEMENTS

This book is a celebration of the 40th anniversary of the second-oldest operating helicopter company in British Columbia, Vancouver Island Helicopters Ltd. (VIH). Our earlier book, *Helicopters: The British Columbia Story*, chronicled the beginnings of helicopter flying in B.C. It was published by the Royal British Columbia Museum and CANAV Books in 1985, and is long out-of-print. This new volume, among other things, brings the story of VIH up to the present and provides new insights into the pioneering years of helicopter flying in western Canada. We are grateful to the Museum – which holds the copyright on our previous collaboration – for allowing us to use select early excerpts from that book, and for generous assistance as we worked on this one – with specific thanks to Bob Griffin and Gerry Truscott.

We also sincerely thank Keith Routley of VIH for suggesting we write this book and the many people, recently and in the past, who have given generously of their time and knowledge, making it possible for us to write with confidence. Since we did not intend to produce a definitive history of helicopter flying in B.C., we have relied heavily on interviews with the participants, backed up, whenever possible, by corroborating evidence from personal log books and similar primary sources. Alf Stringer who, with the late Carl Agar and Barney Bent, founded Okanagan Helicopters Ltd., has given us invaluable help in assembling the early history of that company, as has Barney Bent. With VIH, Lynn Stringer, who founded the company with her first husband, Ted Henson, has been similarly generous in helping us to compile the early history of that company. For the more recent history, we thank father and son, Frank and Ken Norie, and Barrie Hewko (who have read the manuscript for us), Charlie Mooney, Bob Reimer, Larry Huff, Bill Buhr and Trevor Deeley, As well, we are grateful to the following for information, photographs, and all kinds of other assistance: Ann Agar,

Gordon Askin, Gordon Bates, Ken Blackwood, Lori Boechler, Carl Brady, Bill Brookes, Jacquie Burton, Joan Cannon, Ada Carlson, Murray Conway, Eric Cowden, Ian Duncan, Jock Graham, the late Art Johnson, John King, Lynn Klein, Randy L'Heureux, Mike Long, Hugh Lyons, Bill McLeod, Ernie McMinn, Brian Mycroft, Bruce Payne, Uisdean Ross, Keith Rutledge, Jim Stone and the late Bob Taylor.

A particular thanks to Martin Lynch for his incisive editing and expertise – his involvement lends us an authority we don't always deserve – and to Bob Turner, long-time and invaluable colleague and advisor. As ever, we consider ourselves very fortunate in our publisher, Ann West, printer, Darryl Morriss, and designer, Jim Bennett.

Peter Corley-Smith and Dave Parker
Victoria, B.C., July 1995

The distinguished helicopter designer, Igor Sikorsky (left), with Alf Stringer and Carl Agar. Sikorsky admired their accomplishments, and they became friends.

UNITED TECHNOLOGIES/SIKORSKY

CHAPTER ONE

Beginnings: Okanagan Air Services Ltd.

By 1940, most of the labour, and often hardships, of travel in this difficult province had been alleviated. The backpack, canoe, horse and dog had been succeeded by steam: trains running for the most part east and west and sternwheelers connecting the railway north and south. Then came the cars and trucks – all of which followed valleys whenever possible, climbing only to cross the lowest mountain passes available. Finally, aircraft had made it possible to cross the mountain in any direction. Only one area remained that was almost as difficult to reach as it ever had been: the mountains themselves.

At least two categories of people needed to get right into the mountains. Ore bodies seldom reside conveniently in valley bottoms; consequently, prospectors and geologists still had to slog it out on foot to get into the high country. The other category needed, if possible, to get to the very top of the mountains. They were the surveyors, who laid out triangulation grids – grids from which the cartographers made their maps. The higher the surveyor can get, the longer the sight lines and the more accurate the grids. In retrospect, the helicopter was clearly the answer for both groups.

Oddly enough, though, when helicopters were introduced into the province in 1947, this apparently never occurred to anyone. More than a year passed before the real potential became obvious.

And that's how it started: through Carl [Agar] and Barney Bent and some of the others sitting in a coffee shop in Penticton and reading a magazine.

Alf Stringer, Engineer and president of VIH

The story began in Penticton, where two young men, both of whom had served in the Royal Canadian Air Force during the Second World War, were trying to make a career in commercial aviation. Carl Agar and Barney Bent began, as did so many others at the time, by trying to run a flying club – this one in Penticton. One of their first requirements was to find a competent maintenance engineer to look after their machines, and Agar travelled to Vancouver to persuade Alf Stringer, who had served with them in the Air Force as an engineering officer – but was then working as a garage mechanic in the lower mainland – to join them. Alf didn't really regard it as a golden opportunity. He recalls that:

> Carl Agar came down to Vancouver; he spent some time talking to me and decided, I guess, when he went back to the people who were going to form this club, that I'd be a likely subject to tighten the nuts and bolts for them. They decided to hire me and I headed up to the Okanagan Valley. The deal in Penticton didn't sound too awe-inspiring but there wasn't much going on in aviation at that time anyway, so I thought I'd give it a whirl.

His pessimism was justified because business was slow. Few people at the time could afford to fly for pleasure and there was a surplus of ex-service pilots to fill any jobs that did become available. So the new partners soon decided to branch out into the charter business. They set up as Okanagan Air Services Ltd. Unfortunately, their fleet consisted of two de Havilland Tiger Moths, in which they could transport one passenger or the freight equivalent in the front, open cockpit: clearly an uneconomical payload. Their next move was to look into aerial crop spraying. As a prime fruit-growing area, the Okanagan Valley was home to a multitude of orchards, all of which needed spraying to protect them from the insects and blights during the summer season.

If aerial spraying seemed an obvious commercial possibility, a look at spray operations in the Wenatchee district of Washington State tended to be discouraging. "We were shocked at the mortality rate of their pilots and aircraft," Barney Bent recalls. "We looked around at one another; there was Carl Agar and myself as pilots and Alf Stringer as the engineer. I think it was Alf who summed it up. "Well," he said, "we could last at least six months with you two."

Alf Stringer describes the next move:

> In 1946, Carl and a couple of the people who were involved with the flying club were sitting in the coffee shop in Penticton. I guess Carl had picked up the mail and read in one of the aviation magazines that a company called General Aircraft, which operated from Yakima in Washington and had been in the aerial spray business for years, was going to go into helicopters because they looked like the answer. [The principal advantages were slow speed and remarkable manoeuvrability, and the fact that the downwash from the rotor blades drove the insecticide into the foliage of the plants below.]
>
> We were thinking of equipping one of our Tiger Moths for spraying. So we thought, well, since we didn't know anything about spraying, we might as well start out with the unknown. If we could get some fruit growers interested in helicopters –

There is a claim that a British Columbian achieved the first helicopter flight in a steam-operated machine in 1903. His name was Lou Gagnon, a miner at the Nickel Plate mine in Rossland; and the claim was investigated some 30 years after the event by Father Freney – a journalist for ten years before he became a priest – and found to be persuasive but not conclusive.

It appears to be a hybrid helicopter-autogyro. Apparently the main rotor had fixed pitch and there was no mechanism to tilt the disk. Steering (and presumably counteraction of torque from the main rotor) was achieved by a pivoting vane (rudder) mounted behind a conventional propeller. Fore and aft control was achieved by moving forward or back in the fuselage (no mention was made of gyroscopic effect). Nor is there any mention of lateral control. Nonetheless, witnesses claimed it was only pilot error that caused the crash, after a two- or three-minute flight behind the Great Northern Hotel, that virtually demolished the machine and seriously injured Lou Gagnon.

Lou Gagnon's flying machine as portrayed by Father Freney and various Rossland pioneers

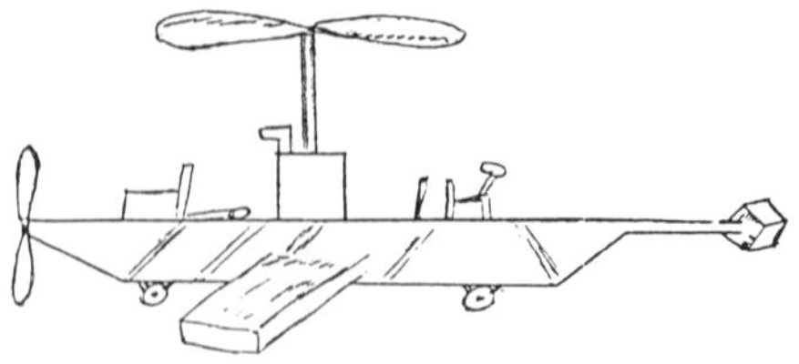

RICHARD COOPER

we knew the fruit growers were interested in spraying – and get some money together, we'd get into the helicopter business.

Unlike Agar and Stringer, Barney Bent had other resources besides flying. His family ran a metal fabrication business, one of the offshoots of which manufactured ground spray equipment for the fruit growers: consequently, he had a natural interest in the potential advantages of using helicopters for this purpose, and he agreed to a preliminary visit to the experimental helicopter operation in Yakima. He, Agar and Stringer went down together, and returned to Penticton convinced that helicopters would provide all the answers to the problems of spraying orchards. As it turned, out they were wrong about the spraying, but fortunately for them the helicopter had a far more expansive potential.

To continue Alf's account:

So we went down to Yakima, where the Bell company had a helicopter. They were actually still testing – they weren't even at FAA [Federal Aviation Administration] certification yet – but they took us up for a demonstration ride. Then we got back and tried to work up some enthusiasm among the fruit growers, then formed a public company and got the helicopter business started. In the spring of 1947, I went down to Yakima and started on a maintenance engineer's course. After a while Carl came along and started on the pilot's course and we went on from there.

However, while the three principals of Okanagan Air Services may have been confident that helicopters were the answer, persuading businessmen to invest was another matter – until, as is the case with many successful business ventures in their early stages, luck intervened. A group of businessmen in Penticton had decided to form a new airline to serve the interior of the province. Led by Douglas Dewar, a distinguished New York chartered accountant who had retired in Penticton, and Ernie Buckerfield, owner of a large feed supply company in Vancouver, along with Ormonde St. P. Aitkens, Gordon Butler and James B. Kidston, they had applied to the Air Transport Board for a licence to operate. Their application was turned down in favour of Canadian Pacific Airlines.

Okanagan Air Services was quick to take advantage of the situation. As Barney put it:

So we said to ourselves, here's a fine opportunity; here are two or three people of substance who have put their financing together for the other airline – let's see if we

> can interest them. Which we did. They weren't exactly overjoyed to get into the helicopter business because they'd had their hearts set on the fixed-wing deal. But Mr. Dewar did agree to head up the company. Even so, we had all sorts of problems raising $50,000 to buy the first helicopter. A helicopter at that time cost $35,000, plus a few spares and some training for Carl Agar and Alf Stringer.

ALF STRINGER

The machine was the Bell 47 B3, powered by a 170-horsepower Franklin engine. Instead of sitting horizontally as it did in a fixed-wing aircraft, the engine was mounted vertically, with the fuel tank beneath and the transmission and rotor mast above. The early models required plenty of maintenance. Grease nipples and grease boots were still in the future, so the helicopter had to be taken apart after every 25 flying hours to be greased. The transmission, too, had to be taken apart and have bearings replaced and hand-packed with grease every 25 hours. The maintenance engineer had his hands full.

Fortunately for Alf, he was trained by an expert. Joe Beebe of Central Aircraft in Yakima had worked with the Bell from early design stages during the war; consequently, he knew as much about the Bell 47 as any engineer. For Carl, things were different. He was trained by Carl Brady, who had only 25 hours of helicopter flying time. So the two Carls were very much learning together.

When they actually began operating in British Columbia, everything looked promising. Dr. James Marshall, an entomologist at the Dominion Experimental Station in Summerland, gave considerable support because he was aware that the fruit growers in the Okanagan were in danger of unintentionally sterilizing their land with too much ground spraying of insecticide. He was confident they could accomplish the same results using a fraction of the amount by spraying from the air. He was right, yet there were many setbacks before they discovered this. The first operational flight took place, with insecticidal dust, over a test plot on the Wally Mutch orchard, just outside Penticton. The downwash from the rotor blades bounced off the ground and the helicopter, with its open cockpit, promptly disappeared from sight. Liquid spray equipment was installed for the next test.

Now they experienced problems with the spray equipment and the logistics involved. Alf found himself spending far more time clearing plugged spray nozzles than on maintaining the helicopter – and when they did manage to keep the spray equipment operating for any length of time, they made the second disconcerting discovery: they could spray at the rate of an acre a minute.

This would have been enormously encouraging if they had been dealing with large acreages; but the orchards were small and diverse and each required a different mix of insecticide. Had they fully succeeded in solving the spray equipment problem, they would have needed a dozen helicopters all working at once to meet the demand during the relatively short growing season. Ironically, their very success made the operation unmanageable, for they would also have needed a large staff, equipped with a fleet of fuel and chemical tankers and other vehicles to support the helicopters. Then there would be the problem of what to do with the staff and equipment when the spraying season was over.

Just as they were facing these problems, disaster struck. Carl flew into some power lines and seriously damaged the helicopter – though fortunately not himself. It was a discouraging moment for the fledgling company, but apparently their backers never wavered. As Barney recalls: "We had a quick meeting right after it happened and Mr. Dewar and Mr. Buckerfield put it to us: 'Do you want to carry on or don't you?' As far as I was concerned, it was up to Carl – he was flying the thing – and there was no question in his mind. 'Get the thing fixed; let's go back at it,' he replied, and that's just what we did."

They had started work at the beginning of August with their first machine, CF-FZX. Now, less than a month later, it was grounded – damaged beyond immediate repair. However, there was still two or three weeks of contract spraying to be done and they managed to lease a Bell 47 B3 helicopter, CF-FZN, from Skyways Services of Winnipeg. Skyways had decided to open up a base in Vancouver but couldn't find any work for the machine, so it was available. In the fall, after completing their first spray contract with the leased machine, Alf put FZX on a truck and drove it to Bell's facilities in Yakima. He spent most of the winter rebuilding it from scratch, the first of many helicopters he was to repair in the years to come.

In the spring of 1948, Okanagan learned another lesson about helicopters that, in retrospect, seems obvious. While a majority of the population might suffer from natural disasters, helicopter operators almost invariably benefit from them. When floods, landslides, avalanches or forest fires occur, the

Carl Agar was not the only pilot on survey work in 1948. Al Soutar, of Kenting Aviation Ltd., flew a summer contract with CF-FJA in the Watson Lake–Teslin area of the Yukon for the Dominion Geodetic Survey (*Canadian Aviation*, March, 1949). Al Soutar reported a successful operation of 210 flying hours, hampered by bad weather. However, F. P. Seers, of the Federal Geodetic Survey, reported that the operation was not a success, "due to extremely unfavourable weather and the very limited payload of the helicopter above 4,000 feet [1 200 m]."

The Canadian Surveyor, July 1949

Carl Agar in the Bell 47 B3, CF-FZX, on the outskirts of Penticton, August, 1947.

OKANAGAN HELICOPTERS COLLECTION

The closed cockpit version of the Bell 47B in use at Yakima when Carl Brady trained Carl Agar in July 1947.

CARL BRADY COLLECTION

FZX with Carl Agar (left) and Alf Stringer on the day they arrived back in Penticton after the ferry flight from Yakima, Washington, July, 1947.

STOCKS FAMILY COLLECTION/
HERITAGE PHOTO-CO-OP

Landing pads, like this one on a side hill, allowed the pilot to do a jump take off and dive down the hill to pick up airspeed.

BILL MCLEOD

Surveyor Gerry Emerson, Topographic Surveys' party chief for its first helicopter operation, refuelling FZX.

B.C. MINISTRY OF ENVIRONMENT

The early machines required a good deal of maintenance. Here, an A-frame is being used to remove the mast and blades so that the bearings in the transmission can be replaced. This had to be done after every 25 flying hours.

OKANAGAN HELICOPTERS COLLECTION

A little reminiscent of a flying bedstead, FZX drops down for a landing.

VERNON MUSEUM AND ARCHIVES

A Vancouver Water Board employee loading insulation and other supplies onto the helicopter bound for the Palisade Lake Dam site.

OKANAGAN HELICOPTERS COLLECTION

Late in August, 1947, Carl Agar flew into a power line. FZX had to be taken down to Yakima on a flatbed truck and rebuilt. Meanwhile, FZN was leased from Skyways of Winnipeg to complete the season's contracts.

OKANAGAN HELICOPTERS COLLECTION

This 1949 photograph of a Topographic Surveys crew was taken north of Hazelton in the Kispiox River Valley. Carl Agar is seated directly below the rotor mast. To his left are the cook and Frank Speed. Gerry Emerson is standing behind Speed; the others were local people hired for the job.

B.C. MINISTRY OF ENVIRONMENT

Ernie McMinn, shown here on the side rack of a Bell 47 D1, was involved in helicopter operations of the Topographic Surveys Branch from the very beginning.

B.C. MINISTRY OF ENVIRONMENT

The evolution of the 47 B3 to the D1 was rapid. Here the covering has been removed from the tail boom of FZN, a "harp" has replaced the tail skid and skids have replaced the wheels. Soon after this the fuel tank was moved up behind the rotor mast and a complete bubble was installed.

BILL MCLEOD

Field maintenance. FZX with the hub and blades off and with four busy engineers at work.

Barney Bent, one of the founders of Okanagan Helicopters, served as a flying instructor with Carl Agar in the RCAF.

OKANAGAN HELICOPTERS COLLECTION

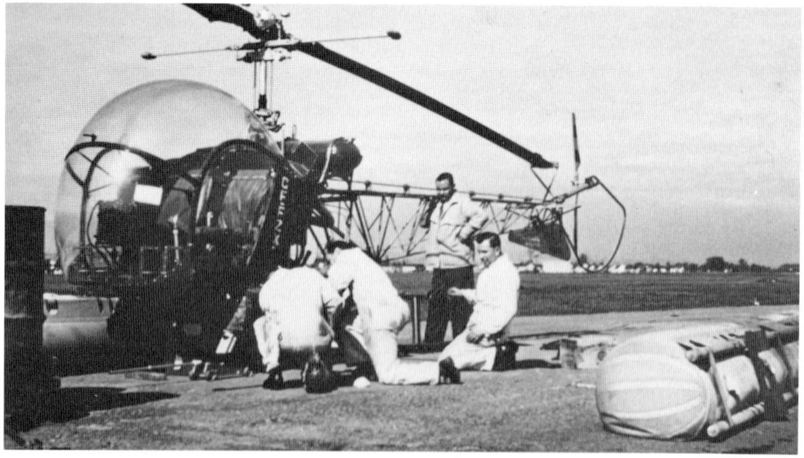

Exchanging floats for skids on the final transformation of FZX from a B3 to a D1.

ALF STRINGER

CF-GHV, the first Sikorsky S-55 acquired by Alcan, was operated by Okanagan Helicopters. Carl Agar, on the ground, is chatting with Bill McLeod after the ferry flight from the Sikorsky plant at Bridgeport, Connecticut, to Vancouver.

ALCAN COLLECTION

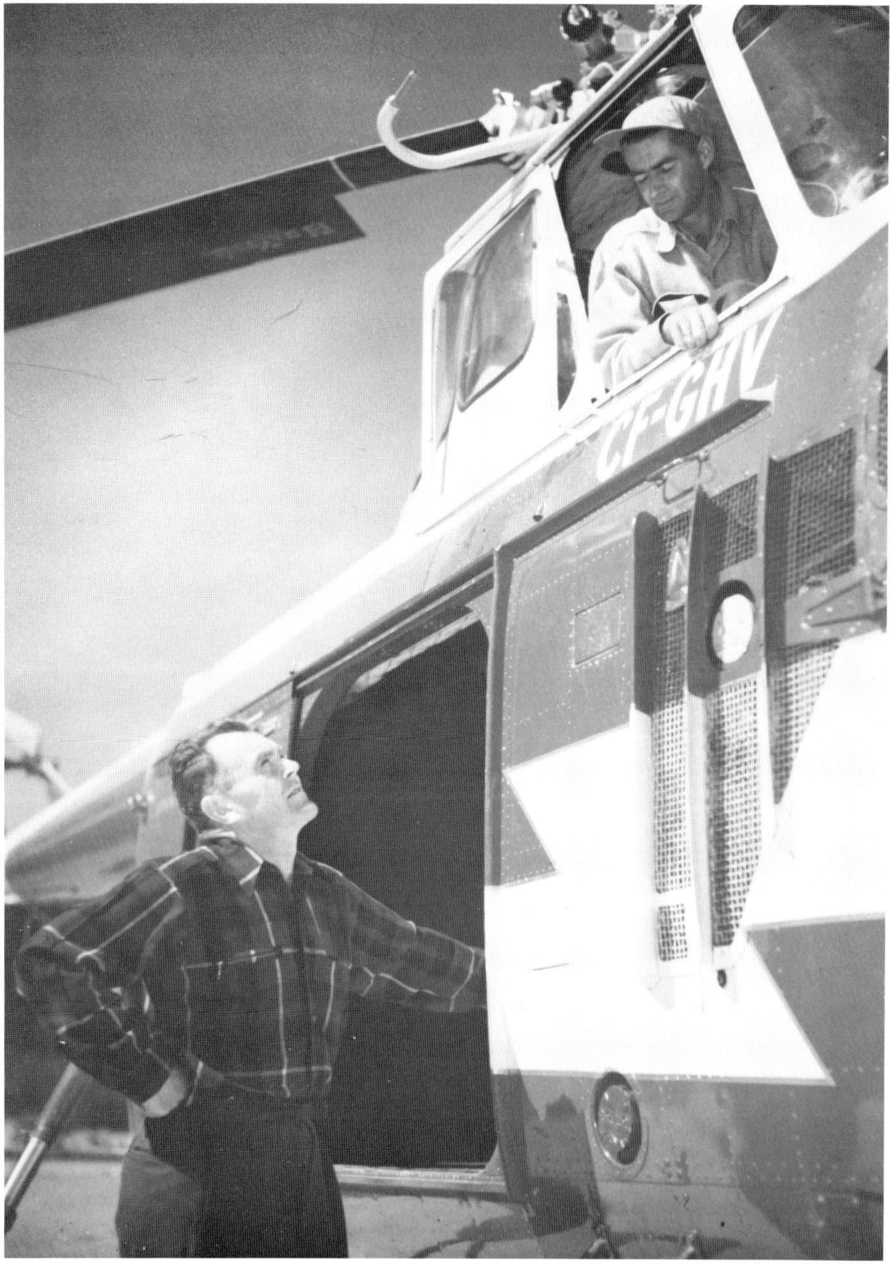

The S-55 gave a big boost to production on the Kitimat–Kemano project. Its capacity to haul 1500 pounds [680 kg] on the hook was more than three times the capacity of the 47 D1s.

ALCAN COLLECTION

One of Okanagan's 47 D1s, CF-GGC, on Alcan's Kitimat–Kemano project, 1953. ALCAN COLLECTION

BOB TURNER

An irreversible.

First commercial helicopter operations in Canada:

1. March 1947, Photographic Surveys Ltd., Toronto, Ontario, Bell 47 B3, CF-FJA, pilot Al Soutar.
2. June 1947, Skyways Services Ltd., Winnipeg, Manitoba, two Bell 47 B3s, CF-FQR and CF-FQS, pilots Paul Ostrander and Jim Sampson.
3. July 1947, Okanagan Air Services Ltd., Penticton, B.C., Bell 47 B3, CF-FZX, pilot Carl Agar.

helicopter becomes invaluable. The spring of 1948 brought one of the biggest floods in the history of the lower Fraser Valley. Among the consequences, the flood left behind it a multitude of ponds and sloughs – prime breeding grounds for mosquitoes – a plague of them severe enough to endanger people's health. Okanagan Air Services was awarded a government contract to spray parts of the valley. This turned out to be more practical than orchard spraying. Carl was faced with wide open fields and sloughs, and the spray consisted of 99 percent fuel oil laced with DDT, a mixture that did not clog the jets of their equipment. The operation was a success.

Meanwhile, Okanagan had doubled its fleet. The machine it had leased from Skyways the previous summer had returned to Winnipeg at the end of the season and it too was wrecked in an accident. Skyways ran into financial difficulties, was unable to rebuild the machine and went out of business. After some skirmishing with the insurance company, Okanagan took over the remains for, in Barney's words, "a dollar down a dollar when they caught us sort of thing. The insurance company was happy to make almost any kind of a deal to get it off their hands."

The spraying job in the Fraser Valley had attracted attention and publicity, and this led to another worthwhile operation – a contract with the B.C. Forest Service to spray in the Windermere Valley, in the East Kootenays, where the forests were infested with the "false hemlock looper" (*Nepytia canosaria*). This operation lasted from July 1 to 24, and it was easy spraying: the only potential problem was that they were working between 2,500 and 4,500 feet (760-1 370 m) – altitudes which even the maker of the helicopter considered marginal for operations because, as you go higher, the air becomes less dense, the rotors provide less lift and engine power diminishes.

Carl Agar, however, was more than equal to the challenge, and began to develop the mountain-flying techniques for which he later became celebrated. He found that flying over mountain ridges and landing in small clearings or on sandbars in the creeks and rivers was not only feasible but thoroughly exhilarating; and perhaps what was more important, the Forest Service was delighted with the results. For the first time, everyone involved began to realize the potential of the helicopter for other tasks beside spraying. People and equipment could be lifted over mountains and dropped in small clearings, perhaps even dropped on mountain tops. The future looked promising if they could make their machine do what Carl Agar felt it could do.

Carl still had much to learn. Just as, when driving a car, the driver has to rehearse certain motions, thinking consciously about how to change gear – right foot off the accelerator, left foot down on the clutch, move the gearshift, left foot up, right foot down again – and so on. In time, of course, this becomes a reflex. The difficulty faced by early helicopter pilots was that they had to unlearn the reflexes they had developed in flying fixed-wing aircraft and learn a new set of reflexes. In a fixed-wing aircraft, you must maintain forward speed because it is the relative airflow, acting on the wings, that provides lift. If you let the speed drop below a certain level, you will stall and drop out of the sky. In a helicopter, because the wing is rotating at a more or less constant speed, you do not have to worry so much about forward speed to stay in the air, unless you are high in the mountains where the air is thin.

In practice, though, forward speed does provide more lift to a helicopter, and this was the problem Carl Agar had to solve. He had discovered the difficulties of getting into the air on hot days even at relatively low altitude. The trick was to lift off and then move rapidly into what is called translational lift: the extra airflow, and thus lift, generated by forward speed. In short, you had far more lift once you had achieved 30 or 40 mph (50-65 km/h), than you had when hovering. It wasn't an easy problem to solve because in physics there are no miracles, and the helicopter pilot has another anxiety. He has to maintain rotor speed – a speed which decays quickly if too much load is put on the rotary wings.

On the early Bell helicopters, the engines had to operate at between 2,900 and 3,100 revolutions per minute, the rotor at about one tenth of this speed – from 290 to 310 revs. The "collective" lever that controlled this aspect of helicopter flying lay to the left of the pilot's seat. Moving it up and down altered the pitch on the main rotors, giving more or less lift, and the twist-grip throttle on the end of the collective fed in more or less power to maintain the desired revs.

Another difference between the aeroplane and the helicopter is that the wing on an aeroplane is relatively rigid and supports the weight of the aircraft by its structural strength, whereas the helicopter's blades, which are wings, are much more flexible, relying on the centrifugal force imparted by rotation to support the weight of the aircraft. Consequently, if the revs drop too low, the rotor blades will simply fold up like the spokes of an umbrella in a gale,

During a flood on Horetzky Creek, two workers on a bulldozer that had dropped into a hole were left standing on top of the bulldozer, waist-high in the water. Bill McLeod took the doors off his helicopter and went to rescue them:

"So I went over and I hovered above the bulldozer and I started coming down to lift the first man off. When I was still a foot and a half in the air, one of the two guys made a wild leap onto the front of the skid. The machine dropped quite violently and the nose of the skids actually went into the water. So I pulled back hard on the collective and by golly, I was still turning to the right. Nothing on the right except trees, and that's where I wound up. . . . When the dust settled, I was still 30 feet [9 m] off the ground, inverted, in the trees – swaying gently up and down, listening to the pitter-patter of the raindrops."

Bill McLeod, Pilot

◁ The transmission, with the swash plate above and irreversible to the right: Bell 47 D1.

and the helicopter will fall out of the sky. At the other end of the scale, if the revs are allowed to climb too high, the blades are likely to break loose from too much centrifugal force, with a similarly undesirable consequence.

Another thing Carl learned was that, if the helicopter was lifted off the ground with some vigour, even on hot days the momentum in the rotor system would keep the revs up for some seconds before it began to decay. After thinking about this and the problem of high-altitude landings and take-offs, Carl came to a conclusion: he knew that if he came into a high landing with some airspeed, killing forward motion just as he reached the landing spot, he could land – at least as high as 5,000 feet (1 525 m). When it came to the take-off, momentum alone wouldn't be enough to keep in the air long enough to attain translational lift. The answer was to land at the edge of a cliff, or very steep drop-off. Then, when it came to the take-off, jump the machine over the edge and push the cyclic* forward to pick up airspeed in a dive and fly to the next landing.

All this, like most other good ideas in retrospect, required a good deal of intelligence to work out, and must have taken a great deal of courage to put into practice for the first time. Fortunately, Carl possessed both attributes and, by the time Okanagan Air Services had been approached by the provincial government – the Surveys Branch of the Land Department – to experiment with the helicopter in the mountains, Carl was ready to demonstrate that it could be done, even though he still had much to learn.

At the time, topographical survey crews were engaged in triangulation. Using theodolites to measure angles – and consequently establish distances – they were laying out a survey grid from which the cartographers could make maps. In triangulation, the longer the base line the more accurate the grid and the less work has to be done to complete it. Obviously, the higher the surveyors could place their instruments, the longer the base lines. But getting high up with all their equipment was backbreaking work on foot; so they were naturally interested in any vehicle that might ease the burden.

Gerry Emerson, with the support of the Surveyor-General, Norman Stewart, contacted Okanagan Air Services and arranged a contract for 25 hours of flying time during the late summer of 1948. They were to work on a map sheet in the Chilliwack area – in the Wahleach Range which runs southeast towards Chilliwack Lake. The cost was to be $85 an hour for the helicopter. It is interesting to note that Carl Agar was earning only $350 a month for this

* Movement on the horizontal plane – forwards, backwards and sideways – was controlled by the conventional control column between the pilots knees. Called the "cyclic," it provided differential pitch to the blades through a mechanism known as the swash plate. If the pitch of the blade sweeping in front of the bubble was decreased, and the converse occurred on the blade sweeping past the tail, the machine would tilt forward and begin to move forward, as it would in any direction if this differential was applied. This same differential pitch mechanism also solved the most enduring problem faced by early helicopter designers. If the blades maintained equal pitch when the machine was in forward flight, the one moving forward into the airflow would provide far more lift than the one retreating into the airflow and the helicopter would roll over. The swash plate, once again, maintained differential pitch to overcome this problem.

dangerous experimental flying and Stringer only $300 for maintaining the machine.

"Carl Agar and Alf Stringer arrived with their machine at Chilliwack airport," Ernie McMinn, one of the senior surveyors involved, recalls, "and we went down and looked at it. Gerry showed Carl on the map where he wanted to go. He wanted to land on a mountain called Cheam Peak. Carl went off on his own to have a look. When he got back, he said, 'Fine, I found a spot back there and landed on it. Had to chase a black bear off it first, but it's no problem.'"

The spot he had chosen was a small, frozen lake in a cirque at the 5,300-foot (1 615-m) level. It was not an ideal one because he was anxious about the strength of the ice at that time of the year, but on wheels, it provided a temptingly flat surface. As it turned out the ice, which was sheltered from the sun most of the year, bore his weight.

Norman Stewart was his first passenger. They landed on the ice again but decided that it was not reliable for repeated landings; so they moved rocks to clear a flat space on the ridge in front of the lake. Carl left Stewart there and went back for Gerry Emerson and Ernie McMinn. As Ernie describes it:

> We got out on the ridge and had to climb about half a mile to get to the mountaintop. In fact, as we realized later, the ridge Carl had chosen was far from ideal for a helicopter landing. It was at the foot of a glacial cirque – a ridge of gravel in front of the little lake with a wall of ice behind it; and, of course, the cold air was coming down off the ice like a waterfall. Not a good place for any kind of flying machine. Anyway, we climbed the mountain and . . . got back to the landing and Carl picked us up just as the fog was rolling in.

Thus began the first of many helicopter survey operations in which Ernie McMinn was involved. He remembers being excited rather than nervous on those experimental flights. "We thought it was a great thrill, sailing over the ground at 50 miles an hour [80 km/h]. We could see how long it would take to walk through it – if you could walk a mile an hour you were doing well."

Ironically, the biggest problem they encountered in the early operations was with the wheels. When a new form of transport is conceived, there is always a temptation to carry over from the old; consequently, while wheels were obviously unnecessary on an aircraft that could rise vertically and move in any direction, it was still equipped with wheels. These made it hard to find

In 1947, on my first survey job up in Terrace, we had a university student on the crew, and we'd gone out to do a little practice mountain climbing to see how the crew could handle it. We got up on top of a hill and I said, "My God, it's taken us five hours to get here and look at it – you could land a bomber up here." The kid looked around. He said, "Well, why don't you use one of these new-fangled things, a helicopter?" We all laughed about it but. . . .

Ernie McMinn, Surveyor

We have just concluded a contract on a twelve-month basis with the B.C. Electric Company and the Greater Vancouver Water District; for the former we will make a monthly patrol of their power lines, and for the latter we expect to move 500,000 pounds [226 800 kg] of construction material to their new dam site at the head of Capilano.

Douglas Dewar, Managing Director, 1949

There is one report that an RCAF Sikorsky S-51 was used for survey work in the Yukon by the Dominion Geodetic Survey in 1949, ("Report on Operation of Helicopter for Survey," Department of National Defence, 112.3e1 [D32]), but no further information has been found. In 1950, probably spurred by the enthusiasm of the British Columbia Topographic Survey people about their operation for the Bowser Lake map sheet in 1949, the federal government used four helicopters on their operations. Party Chief Howard Spence used two Spartan Air Services Bell 47 D1s in northern Quebec, and A. C. Tuttle used two Kenting Aviation Hiller 360s in the Yukon (*Rotary Winged Aircraft*, nd). From then on, the Federal Department of Mapping and Surveys made extensive use of helicopters on their survey operations.

It used to be that one surveying party in one season would do the control work for one map sheet – one 50,000 map sheet – and Gerry [Emerson] that first summer, using helicopters, did ten map sheets. It was that good: you could get ten times the production.

Ernie McMinn, Surveyor

a landing that was perfectly flat and, since there were no brakes, the helicopter tended to run away down the slope. Alf Stringer provided a temporary solution by jamming strips of old car tires between the strut and the wheel. Bell responded in 1949 by replacing the wheels with the now common skid gear.

Carl went straight from this experiment to another one; he put in 12 hours of timber cruising and the customer was almost ecstatic about the results: the ease with which they could pick out routes for the cruisers to travel on the ground, and the simplicity of deciding where logging roads should be constructed.

The following year, 1949, was a pivotal one for the company. The principals had made the decision to abandon spraying and concentrate on replacing the horse – the packhorse trains that had always been used in the summer to transport people and supplies into remote areas for surveying, mining exploration and timber cruising. To win this sort of contract, they decided they needed to move closer to the seats of industry and government: Vancouver for the one, Victoria for the other. They settled for the former, renting surplus hangar space from Queen Charlotte Airlines at the airport early in the spring.

The result was a series of small but highly visible contracts. The first was to fly supplies for the building of the Palisade Lake Dam in the mountains north of Vancouver, which was to augment the city's water supply. This contract – to move 500,000 pounds (226 800 kg) of construction material – was particularly valuable because it was such a visible operation. People in Greater Vancouver could see the helicopter at work. At least some of them must have begun to realize its potential in their own fields. Carl Agar had hired one of the two pilots from the defunct Skyways company, Paul Ostrander, to fly this contract, while he took the other ones – a mining exploration job, operating out of Tofino on the west coast of Vancouver Island, and another spell of timber cruising north of Squamish – before flying to Lytton, where he put down two geologists, McKenzie and Warren, together with their camping gear, close to the 8,000-foot (2 440-m) level of Gott Peak, leaving them to walk out.

While in Lytton, Carl received a message to return to Vancouver to meet the most important customer he was ever to meet – Wilfred Heslop, a professor of civil engineering at the University of British Columbia, who was acting as a consultant to B.C. International Engineering. Carl took him up in the helicopter for demonstrations in and around the north shore mountains. Heslop had little to say at the time, but he asked Carl to meet him in Terrace on August 31 for two or three weeks of flying.

Carl now returned to fly for the Topographic Survey people in the Nass Valley, after which he joined Heslop. With him, he flew to the then small village of Kitimat, where a stock of fuel and oil had been put in by sea earlier in the year. Although he didn't know it at the time, this was the beginning of the Aluminum Company of Canada's Kitimat-Kemano project which, if not the largest, was certainly one of the largest industrial development projects undertaken in British Columbia. It was also the turning point both for Okanagan and the entire commercial helicopter industry.

Bill McLeod, the first pilot trained by Carl Agar, and engineer Jock Graham arrived on the job early in 1951; Bill describes the beginnings:

> When we went in, my first commitment was for eight weeks. . . . We were there to move the survey crews. They were going in only two weeks ahead of the road-construction crews to build 10 miles [16 km] of road from tidewater on the Gardner Canal up to the actual powerhouse site on the Kemano River. . . .
>
> I found that I was not only flying the surveyors, but all the construction people as well. At the end of eight weeks, you couldn't have got that helicopter out of there with a bulldozer. They'd have killed you if you'd tried to take the helicopter away from them.

By the end of 1951, Okanagan had three of their Bell helicopters in Kemano, and Alcan bought two more to be operated by Okanagan. By April 1954, Alcan also acquired a much larger helicopter: a Sikorsky S-55, capable of carrying a 1,500-pound (680-kg) load, compared to 450 (205 kg) for the Bells; and Alcan added a second S-55 to the fleet before the end of the year. Meanwhile, Okanagan was busy, almost frantically busy, hiring and training more crews. The company now had the security of year-round operations out of heated hangars, as well as a steady and predictable income. In short, Okanagan, which had changed its name to Okanagan Helicopters Ltd. in 1952, was on its way to becoming one of the largest helicopter operations in the world. By 1955, the company was employing 90 people and its helicopters logged 12,000 hours of revenue flying–enough to provide handsome profits.

I landed on a ledge at about 6,500 feet [1 980 m]–a short cliff on one side and a sheer drop on the other. The ledge was about 150 feet [45 m] wide and I landed facing the cliff, about 60 feet [18 m] from the drop-off. The only way to get off was to over-rev and drop off the edge, then dive down to pick up speed and get the revs back up. But I was green then and didn't land close to the edge. I had to jump, turn around and go for it. I knew right away that the tail wouldn't clear. I shoved hard forward on the stick and kicked on rudder, cart-wheeled over and I was inverted at one point; then pulled round and got away with it. You had to learn the hard way: never land far from a good drop-off and land parallel to it so that you can jump the machine sideways and protect the tail rotor.

Bill McLeod, Pilot

Snow could be tricky. If it had a crust on it, you could break through and drop a foot or two–enough to ding your tail rotor. So we fitted bear paws: pieces of plywood about 30 inches long and 15 inches wide (75×40 cm]. They would be clamped to a piece of tube and bolted to the back of the skids. Then, if the machine settled, the nose would go down, but the bear paws kept the tail rotor out of trouble.

Bill McLeod, Pilot

CHAPTER TWO

Competition: Vancouver Island Helicopters Limited

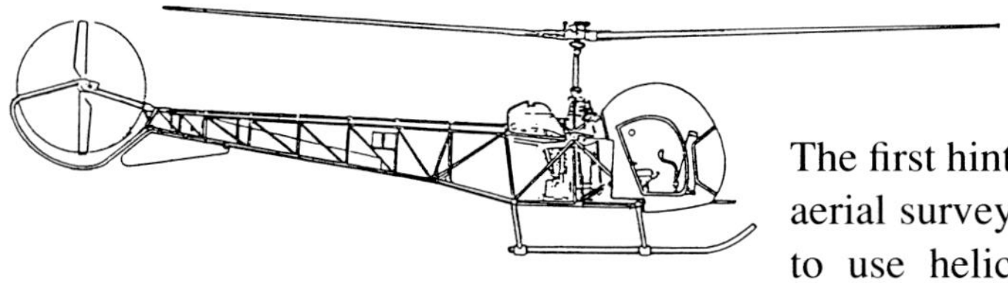

It was hairy of course. You never landed far from the edge of a drop-off. You just had to hop, skip, get over the edge and clear your tail rotor and get some rpm on the way down. Terrifying.

Bob Taylor, Pilot and Operations Manager

The first hints of competition to Okanagan Helicopters appeared in 1950. An aerial survey company in Ottawa, Spartan Air Services Limited, had started to use helicopters to provide the ground control – the surveyed heights and bench marks – it needed for mapping projects. Spartan, like Okanagan, expanded rapidly. By 1955 it had a fleet of more than 12 Bell 47s; two years later it had a base in Calgary and was operating across the country.

Closer to home, on Vancouver Island, a young man named Ted Henson was watching the success story across the Strait of Georgia with increasing interest. Ted had been in the RCAF, hoping to become a pilot, but joined too late in the Second World War to achieve this ambition. Returning to Victoria after his discharge, he began work as an automobile mechanic. Then he suffered a bout of pleurisy, and while being treated in the Royal Jubilee Hospital met and fell in love with Lynn, a trainee nurse. Lynn abandoned her nursing career to marry Ted and went to work as a dental assistant, while Ted, with the help of a friend, Ron Page, managed to finance the purchase of a small Ercoupe monoplane in which he learned to fly and eventually earned his commercial pilot's licence. As it turned out, Ted did very little commercial flying in fixed-wing aircraft.

Times were hard for Ted, just as they had been for Carl Agar and Alf Stringer when they were trying to get back into aviation after the war. When he conceived the idea of getting into the helicopter business the problems of financing seemed insurmountable.

He began by writing the Bell helicopter people to tell them about his ambitions. Bell responded by putting him in touch with the company's western representative in Los Angeles, Art Fornoff. Fornoff came up to Victoria to see Ted. Apparently satisfied that Ted was both sincere and capable, he arranged for an interview in Seattle between Ted and Bill Boeing

Jr., son of the founder of one of the largest aircraft manufacturers in the world, who was forming his own helicopter company, Aero Copters Inc.

Fortunately Bill Boeing and Ted hit it off and the immediate outcome was the formation of a partnership. At the time – this was in 1953 – an American could hold only a 40 per cent interest in a Canadian company. Boeing's share of the company was to pay for Ted's helicopter pilot training and finance the purchase of the first machine, a Bell 47 G with a down payment of $18,000. It was registered CF-IDX. Ted named the company Vancouver Island Helicopters Limited.

All this had been accomplished with relative ease. Now began the hard part – acquiring a licence to operate from the Department of Transport, and the first revenue contracts. Lynn Henson (later Lynn Stringer) describes the problems:

> It was the biggest battle of our whole life; it was so hard. You have to appreciate that we had no money. In September [1954] Ted had to quit his job as a mechanic because he had all this work with the business. The Liberals were in then and it was very difficult to get a licence. You had to prove that the traffic was there and that there was a public need for helicopter service – which we did because we had done our homework. Then, when he put his application in, Ted made the mistake of asking for a protected base. This meant that no other operator could come within a 25-mile [40-km] radius – and Okanagan fought that; they fought it very, very hard. Things were really tough for us. Ted had quit his job and my pay was very small. There we were in a little tiny house, 700 square feet [65 m^2], but with a big garden that kept us going.

Okanagan's objection to the founding of VIH had come as an unpleasant surprise, and while the Hensons waited with growing anxiety to see what the outcome of the objection at the Air Transport Board hearing would be, Ted was busy trying to raise capital to run the company. He went practically from door to door, urging people to invest, usually with indifferent success. The first principals, apart from Ted and Bill Boeing, were Ron Grant, a lawyer, and Dave Turnbull, a businessman whom Ted had met at the Victoria Flying Club.

At the same time, Ted was busy trying to find prospective customers. The B.C. Forest Service and B.C. Power Commission were his main targets. They appeared interested but they would not commit themselves. He received some encouragement from the newspapers, which followed his story with considerable coverage.

My brother, who was very interested in helicopters, was given a 15-minute ride as a present. That was about the minimum time you could go for and it cost about $25; that seemed like a lot of money in 1959 or 60. Bob Taylor was the pilot and he showed us all over the Saanich Peninsula and down over our home in Victoria. He explained all about the Bell and showed us how everything worked. It was a great flight. When we got back he smiled about the long flight and said not to worry, that he figured he should check out the helicopter before it was sent off on a contract. Then, he asked our parents if they would like a short flight, which they accepted very quickly. I always remember how generous and encouraging he was to a couple of interested kids. All we were expecting was a couple of circuits of the airport.

Bob Turner, Historian

One geologist had a ten-foot [3 m] inflatable dinghy. He wanted to be dropped by a lake so that he could paddle around it doing his geology; and then moved to the next lake. For obvious reasons, he didn't want to deflate it and have to re-inflate it every time. We tried tying it on the starboard float, on its side. I couldn't see anything on that side, but I reckoned I could manage. The wind got inside it and I ended up travelling sideways. The next time, I tried hanging it on the hook with a 30-foot [9-m] line. After climbing away, we settled down to cruise at about 50 mph [80 km/h]; then I looked over to the left and found the dinghy flying along beside me.

Des O'Halloran had the same problems in 1961.

Peter Corley-Smith, Pilot

DES O'HALLORAN

Okanagan's objections were carefully phrased. There simply wasn't enough business to justify another helicopter company; it was bound to fail. "We are not," Okanagan assured the Board of Transport Commissioners, "apprehensive of competition. Actually we welcome competition if it is shown to be in the public interest."

To which Ron Grant, Vancouver Island Helicopters lawyer, responded dryly: "It is unusual for a business concern to set itself up as an economic watchdog."

The newspapers also referred to a mysterious American partner in the business (Bill Boeing, presumably), hinting at a threat to economic nationalism. At the same time they ran photographs of Ted and the brand new machine, CF-IDX, reporting that he was operating out of a field on Shelbourne Street. In fact, as Lynn points out, such operating as Ted was doing at the time was from the Howroyd strip in central Saanich. Joe Howroyd, always active in Victoria flying circles, was a friend and the landing strip was a very convenient base, at least for the moment.

In April of 1955, the long awaited letter arrived from the Department of Transport informing them that their application for an operating licence had been successful. Now Ted had to move from the casual phase of parking the helicopter on a friend's landing strip into some sort of hangar where proper maintenance could be performed. Lynn describes this move:

> Ted was looking for a suitable place, with some room for an office, and he went out to the airport and here was this war-surplus hangar, packed high with old windows. They were using it as a warehouse. Every window on the hangar itself was broken. Just to look at it was so disheartening. But what else was there? So he said he'd take it. I think he paid $132 a month. He had to replace all the broken windows; then he cut a hole in the front of the hangar for a door so the helicopter could get in. There was just room for one helicopter and a little office.

Nevertheless, they were in business and Ted left in May, with his mere 25 hours of helicopter flying experience, to work for the B.C. Power Commission on the Homathko River, one of the roughest sections of terrain on the B.C. mainland coast. Although the Power Commission was examining the potential for a dam and a generating plant on the Homathko, the actual work with the helicopter that first year was done for the Power Commission by the Surveys and Mapping Branch of the provincial government. The engineering and environmental problems apparently could not be solved and no dam or generating plant was ever constructed, but the exploratory work continued for three years.

Ted was accompanied on this operation by Ted Prothero, a fixed-wing engineer who had just earned his helicopter endorsement after a six-week course with Bell. Then, only a month or so later, Ted Henson suffered an accident, just as Carl Agar had on his first revenue operation: an accident that seriously damaged the machine.

Ted's accident was even more discouraging than Carl's encounter with the power transmission lines. He had a partial engine failure with two passengers on board. The only clearing within reach was a small grass-covered sandbank on the edge of the river. He put the machine down parallel to the river. There was just room for his skids on the bank, and no more. The passengers got out while Ted ran the engine down. He switched it off and joined them for a moment to wait for the rotors to stop turning before he pulled some spark plugs to see if they had been the cause of the problem. As he stood there, the sandbank suddenly crumbled into the shallow water and the helicopter fell sideways with the blades still turning. The machine would require a complete rebuild. It was a remarkable piece of ill fortune.

Ted and his passengers then had to endure a three-day walk down the rugged Homathko River to the head of Bute Inlet, where they were picked up by a fishboat. The wreck was soon salvaged with the help of an Okanagan helicopter, and Ted spent most of the remainder of that first year helping Ted Prothero rebuild it.

The new company's second year of operation, 1956, was more encouraging. Ted went out again to the Homathko, to work with the rebuilt CF-IDX; in addition, he leased two machines, together with pilots, from Aero Copters. The two leased machines were flown by Don Scholberg and John Watson. Meanwhile, Ted hired a second engineer, Harvey Jones. They had a successful season and the future began to appear rosy enough for Ted to look for pilots to train for the following year's work.

He chose two people from very different backgrounds. Robert P. Taylor was an ex-RCAF officer who had learned to fly late in the war and became an instructor. After the war he attempted to make a living as an artist but, in spite of his talent, he was only just managing to survive. Early in the Korean War, he decided to rejoin the Air Force and spent the next five years training

Well, we did start slinging that year [1957]. Before then we were moving these drill rigs and having a heck of a time; the engine on a BBS-1 drill weighed about 300 to 350 pounds [135 to 160 kg], and the frame about the same. We used to use outriggers to move them. We'd put the drill rig on the right-hand rack, opposite the pilot, and we'd stick a board out the other side ten feet [3 m] and hang a sack of cement on it to balance the load.

Bruce Payne, Pilot

Joe Howroyd took me to the hangar to meet Ted Henson in 1956, and here was this helicopter, and I looked at the thing and said to myself, well that's the flimsiest looking contraption I ever saw in my life. You'd never catch me in one of those things.

Frank Norie, Pilot and President, VIH

NATO students on Harvards at a training school in Calgary. By the end of five years he was becoming bored with the routine and decided he would like to fly helicopters. There were no openings in the service, so he went to see Ted Henson. Bob Taylor remembered that it was not an easy task.

> A lot of civilian operators didn't jump up and down with enthusiasm at the prospect of hiring a military pilot in those days – didn't think they would prove too useful in the bush, I guess; so we more or less had to prove otherwise. And the conversion from fixed-wing to helicopters was much harder than I had expected. For a while I didn't think I was going to make it. The training was comprehensive: we did 50 hours of basic and then 25 hours of mountain training. Earl Thompson of Aero Copters had already done more than 4,000 hours of helicopter flying time by 1956, and in February of the next year he trained me up in Strathcona Park – Great Central Mountain and Golden Hinde were my initiation into mountain flying. It was hairy of course. You never landed far from the edge of a drop-off. You just had to hop, skip, get over the edge and clear your tail rotor and get some rpm on the way down. Terrifying.

The other pilot Ted chose was J. B. (Bruce) Payne. Bruce had worked in the bush, spending a season with the B.C. Forest Service, on the Hart Highway, between Prince George and Dawson Creek; and another with the provincial Topographic Survey in northeastern British Columbia, building towers for triangulation grids. In the winters he had taken flying lessons at his own expense at the Victoria Flying Club. Ted decided to ask the club for a recommendation. Bruce, who had just turned 20, was chosen and he began training at the same time as Bob Taylor.

In May 1957 both had completed training and were signed on as employees. An agreement had been reached that they would start at a salary of $450 a month, working up $600 after three years, at which point their training would be considered paid for. Bob Taylor went with Ted Prothero to the Power Commission job on the Homathko and Bruce's first revenue job was to fly people from the Department of Transport and Trans-Canada Air Lines up Mount Slesse southeast of Chilliwack, where a North Star had crashed, killing all on board.

For the rest of that season, Bruce worked on a number of jobs – some mineral exploration for Falconbridge Nickel at Bugaboo Creek in the East Kootenays; power-line patrols out of Campbell River, on Vancouver Island; and a stint with the provincial Topographic Survey – before travelling to Ontario to fly a short contract with the Department of National Defence.

Mack Chapman of Fisheries was with me. We were counting fish on the Fraser River, around Tête Jeune. I just got low enough to run into these telephone wires strung across the river. They were about as thick as your little finger. Just about the time I saw them I hit them. They caught the bubble about eye level and they didn't break it, they went up over into the mast and then the blades caught them. They made an awful racket.

We couldn't stay up; it slowed the blades right down and we settled down and landed in the water. It was only three or four inches [8-10 cm] deep.

Bruce Payne, Pilot

Ted Henson (right) with engineer Harvey Jones. The photograph was taken the day before Henson's disappearance at Tatlayoko Lake. BOB TAYLOR

Forecast—Clear, little change (Details on Page 2)

The Daily Colonist.

Vancouver Island's Leading Newspaper Since 1858

Telephone—3-4111

NO. 260—NINETY-NINTH YEAR VICTORIA, BRITISH COLUMBIA, FRIDAY, OCTOBER 18, 1957 26 PAGES

Formal Regalia, Casual Chat

Helicopter Crashes in Lake

'Indestructible' Flier Ted Henson Missing

Victorian Feared Drowned

SEARCHERS READY

STILL HOPEFUL

TED HENSON
. . . in his helicopter

Off B.C.

Ship Fuels Subs

EVIDENCE

Newspaper headline when Ted Henson went missing at Tatlayoko Lake. The "Indestructible" obviously refers to his previous mishap on the Homathko River, when he turned up after being reported missing for several days. DAILY COLONIST

Ted Henson (left) with engineer Ted Prothero. LYNN STRINGER COLLECTION

In frosty weather, as much of the helicopter as possible was covered nightly. Here, the engine and transmission are still enclosed but the nylon covers are being removed from the main blades.

BOB TAYLOR

Lynn Henson at work in her office which, until then, had been the parts room in the hangar at Sidney, *circa* 1958.

ALF STRINGER

After Ted Henson's death, Bill Boeing Jr. organized this team to run the company: (left to right) Bruce Payne, chief pilot; Lynn Henson, president; Bob Taylor, operations manager; and Harvey Jones, chief engineer. LYNN STRINGER

Ted Henson, left, Lynn Henson, ▷ with her back to the camera, and Bruce Payne under the rotor head, Port Renfrew, 1956.

JIM RYAN PHOTO, VIH COLLECTION

Once a few trees had been felled, a helicopter could bring in all the materials for a camp, as well as for its own pad.

JEAN ANDRÉ PHOTO, B.C. FOREST SERVICE

American registered N9848, one of Bill Boeing's Aero Copters' 47 Gs, having a load tied on under the mainframe during the Homathko operation, 1957. Soon after, Ted Prothero devised a release mechanism based on a Canso bomb rack to avoid this dangerous practice.

BOB TAYLOR

Engineer Howie Goodall (left) and pilot Des O'Halloran freshen up KNC at the north end of Stuart Lake, 1960.

DES O'HALLORAN

Variations of the 70mm camera boom resembled something more appropriate to science fiction than forestry.

R. EDWARDS PHOTO
BRITISH COLUMBIA PHOTO SERVICE

Alf Stringer, founder and vice-president of Okanagan Helicopters Ltd. and president of Vancouver Island Helicopters Ltd. from 1965 to 1985.

DAVE PARKER

We'd been out for three months, living on bannock, beans and bacon. Now we were ferrying up from Chukachida Lake to Paddy Lake, near Atlin. There was a brand new crew waiting eagerly for us. But Don Poole climbed up to 5,000 feet [1 500 m] and we chugged on right past Paddy and landed in Atlin. "Time for a couple of days off," said Don. We went to lunch. The first thing I ordered was a salad. "No steak?" Don asked. "Well," I says, "I think I'm getting scurvy: my gums are all sore." "Yup. That's why we're here. Mine are too." We lived like peasants in the bush in those days.

Eric Cowden, Engineer

Meanwhile, Bob Taylor found the Homathko region an eye-opener on his first revenue flying job. He was mostly moving survey crews in preparation for potential dam construction. He had his first experience of slinging when he moved a big Kootenay wood-burning stove into a camp on the river. Kootenay ranges were substantial cast-iron units with double heating ovens. They were too heavy to be carried on one rack, so Bob flew his engineer to the landing site, then returned to tie the stove onto a crossbar under the helicopter with a 10-foot length of half-inch rope (3 m-1.25 cm).

When he reached the landing site and lowered the stove gently to the ground, the engineer, armed with a sharp knife, cut the load loose. It was an illegal and, of course, dangerous procedure. Any sort of mechanical problem en route could have been disastrous. Fortunately it was this experience that motivated Ted Prothero to fit up part of a bomb rack from a Canso amphibian as a cargo hook, so that the load could be jettisoned in an emergency.

The third season, 1957, started out successfully for the new company. Ted Henson was quoted in an article in *Canadian Aviation*: "I started this business because I liked to fly. I've only been in it a year and yet this will have to be my last year of flying. We'll be getting a lot bigger, and managing the firm – mainly lining up work to keep the copters busy – will be a full-time job."

Ted spent only one short summer "flying" that managerial desk. Bruce Payne returned from his job in Ontario early in October, just as one of the Seattle pilots, whose six-month Canadian work permit had expired, was obliged to return home from the Power Commission job on the Homathko. Ted didn't have the heart to send Bruce right back out again; instead he went himself to finish the few days of work left. He was flown by floatplane into Tatlayoko Lake, a narrow and very deep body of water which stretches for some 14 miles (22 km) between steep, rocky shores. The following morning, October 17, 1957, Ted took off in his helicopter for a five-minute flight along the lake to another camp. He was never seen again. The people in the camp he had just left reported hearing a heavy thud. The only remains ever found were a seat cushion and an oil can.

The loss of their president so early in the company's existence was a crushing blow. To Lynn Henson it seemed the end of so many dreams:

> The sad part was that we had to go on. I can't remember now, but I think it was two weeks before the search was called off, everything was finished and he was

declared dead. But in that two-week period, every time the phone rang I thought maybe he'd walked out after all and here he was phoning. But because I'd been on the lake in the summer, and knew how treacherous it was, I couldn't really believe that. Even if he had been able to swim ashore, there was no place for him to hang on – it was just straight down and there were no little beaches for him to lie on or anything.

The crisis was one of management and finances. The financial problem was eased by an unexpected bonus. Lynn recalls:

When Ted and I were sitting at home, waiting for our operating licence – when Ted was still working on cars in the evenings to bring in some money – Fred McGregor, of Mutual Life, came along and sold him a $30,000 life insurance policy. Ted never told me about this. He was paying about $25 a month in premiums; and because we had so little money, he fell behind on the premiums. That was how I found out. Ted was away flying when a letter from Mutual arrived. I opened it and it said that if you don't pay by such and such a date, your insurance will no longer be in effect.

I was pretty angry because Ted hadn't told me anything about this. I phoned Fred McGregor and told him so. I was looking after the bills and we couldn't afford it. I would find the policy and tear it up. I said, "Thank you very much" and "goodbye," and I thought that was the end of it. This was in April '57. I wasn't very nice to him. Then came October '57 and Ted was dead; and you know what happened? Mr. McGregor had been paying those premiums himself for all those months. That's what saved the company – Mr. McGregor's generosity. Without that $30,000 I would have had to sell the company to pay succession duties.

So the financial crisis was solved, but there remained the problem of management. Lynn, who had been doing all the bookkeeping and paper work, was the only person who knew the day-to-day details of the operation. Bill Boeing called a meeting of the shareholders and the decision was reached that Lynn should be the president to run the administration of the company and that an operations manager and a chief engineer should be appointed. As the company's first employee, Ted Prothero was obviously a candidate for the position as operations manager; but in the end Boeing recommended Bob Taylor for the job and Prothero became chief engineer.

For Lynn, the next two years were devoted almost entirely to work in the hangar. Since she had a son who was seven and a daughter who was two, the problems were very real. She remembers the period with mixed emotions:

That was a hard time but I had a very good baby sitter. Danny went to school and we worked things out – even though I was out at the hangar 12 hours a day every day. At the time, though, I came in for a lot of criticism from my close friends, as well as from my family. They seemed to think I should stay at home and look after my children even if that meant going on welfare.

Bruce Payne was a natural. Not only did he have all the physical talents: eyesight, depth perception, judgement, coordination – all those good things – but he was an intelligent innovator. He was an awful pilot to follow on a contract because you constantly heard from the crew what wonderful new things Bruce had done.

Peter Corley-Smith, Pilot

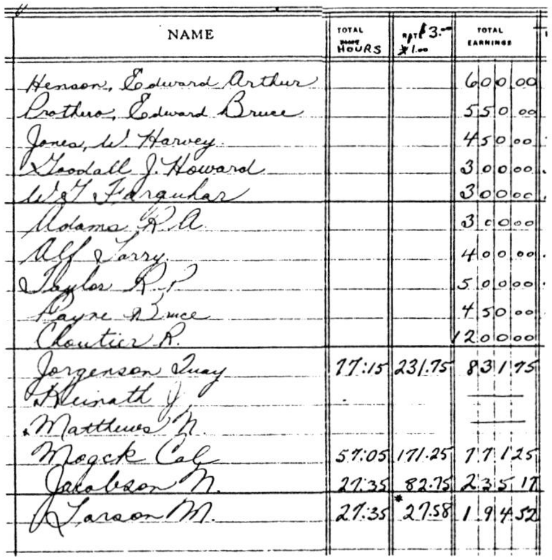

NAME	TOTAL HOURS	@ $3.— *1.00	TOTAL EARNINGS
Henson, Edward Arthur			600.00
Prothero, Edward Bruce			550.00
Jones, W. Harvey			450.00
Goodall J. Howard			300.00
W. J. Farquhar			300.00
Adams R.A.			300.00
Alf Barry			400.00
Taylor R.P.			500.00
Payne Bruce			450.00
Cloutier R.			1200.00
Jorgenson Guay	77:15	231.75	831.75
Kleinath J.			
Matthews M.			
Mogck Cal	57:05	171.25	771.25
Jakobsen N.	27:35	82.75	235.17
Larson M.	27:35	*27.58	194.52

VIH payroll, July 1-15, 1957.

Also, I found out how very quick people are to take advantage of you. One day in a grocery store I met a man who was a friend of Ted's and mine, and he wanted to buy the company. He was into another fixed-wing operation and he wanted to buy our company to amalgamate with his. I said no. He said, "just name your price. Every man has his price." When I refused again he said, "I'll give you six months to run that company into the ground!" I'm proud to think that every day that goes by makes him more wrong.

On the other side of the coin, there were considerable satisfactions for Lynn:

My office and a little room where we kept the spare parts were all we had – and my office was also the pilot and engineer crew room. That's where we all met and I tried to do my bookwork. Mind you, I loved being among them and listening to their talk about their experiences, but I was getting desperate because it was almost impossible to get any work done in that atmosphere. Then one day I remembered the story of Glenn McPherson. His wife wanted a new kitchen and she kept after him and kept after him and one day she took an axe and chopped down all her cupboards. He came home from work and there was this pile of rubble in the middle of the kitchen and she was just sitting there, and she said, "I'm not doing any cooking until I get a new kitchen." So the next day Glen got started and had the kitchen rebuilt.

My own solution wasn't quite so drastic. I just picked up my desk and my typewriter, went into the parts room, plunked down in the middle of the parts and said, "I don't care what you people do; this is where I am going to work from now on." The next day, they started building a new parts room down at the back of the hangar, where it is now. The truth is I loved the work I was doing. I was determined to keep our company going for Ted's sake.

With Bob Taylor running the flying and Ted Prothero the maintenance, business began to flourish. These were still boom years. The Power Commission work on the Homathko continued until 1959; meanwhile, Vancouver Island Helicopters was very active in mining exploration along the Stikine River in northwestern British Columbia. In January 1958, the company had on its books Bob Taylor and Bruce Payne as pilots, Ted Prothero, Ron Adams, Howie Goodall and Harvey Jones as engineers. In February, Des O'Halloran, an ex-RCAF officer, became the company's third full-time pilot. A young Australian, Noel Dodwell, was working as a volunteer in the stockroom. In June 1959 he became an employee at a salary of $200 a month. His perseverance paid off; eventually Bob Taylor trained him as a pilot and he became a very successful one.

This was the year, too, when Bob Taylor landed the company's first B.C. Forest Service contract. Bob recalled that it was a big step forward:

It was the inventory contract, and for us it was a winner from the start. You flew nearly all the time with one passenger, so it was easy on the machine and on the pilot. You had one man on board with a tape recorder and he called what were known as plots – areas of trees perhaps half a mile long marked on aerial photographs – and the "classifier" recorded the species, estimated heights and ages of the trees in each plot as you flew about 50 feet [15 m] above them. Then, after an hour or so, we would land beside a plot right after it had been recorded and the classifier would go in and check his estimates to make sure they weren't too far out.

What made the contract really worthwhile was that it was a big project; the inventory was to cover the whole province over a 10-year span. We had it for five years. Another good thing about it was that the forestry people would let us put a brand new pilot on the job. I'd train him in the winter, and this was an excellent job for him to build up his time without having to try anything too hairy.

Despite Bob Taylor's optimism, the inventory job could be "hairy." At least Noel Dodwell in his second year as a pilot found it so. One of the aspects of the program that Bob did not mention was the use of two-man sampling crews. These crews, almost invariably summer students, were put out by helicopter to sample random plots of trees. They would work out of a fly-camp for several days at a time, using a chain and compass to reach the randomly selected plots, and then physically measure the trees from the ground and record the data.

On this occasion, Noel had dropped the crew's food and camping gear in a clearing on the bank of a creek. Then he took off to drop the two students some three or four thousand feet (900-1 200 m) up the mountain, near the treeline, so that they could put in two samples before walking back down to set up their fly-camp. He couldn't find a landing site where they wanted one and decided to prop one skid on a steep bank of scree, holding the helicopter in a semi-hover while they jumped off.

Before doing this, he explained to them the importance of leaving the helicopter one at a time, and of not lingering on the cargo rack before jumping because he could not hold the helicopter level for more than a few seconds in this unbalanced condition. Satisfied that they understood, he settled the skid on his side of the machine firmly on the scree and nodded to the student on the outside to go. Unfortunately the student was still wearing sneakers; he had tied the laces of his boots together and he was carrying them in his hand. When he got out on the rack and looked down at the jagged scree, some four or five feet (1.5 m) beneath him, it occurred to him that he should

ERIC COWDEN

Noel Dodwell enjoying the amenities of a well appointed bush camp.

be wearing his boots before he jumped, and he calmly sat down on the cargo rack to change into them.

Noel shouted to the student still sitting beside him to tell his companion to get the hell off the rack (or words to that effect). As the second student moved over to do so, Noel finally lost control and the helicopter rolled down the slope, thrashing itself to pieces as it did so. Astonishingly, no one suffered anything worse than minor scratches and bruises; but people in main camp, more than 10 miles (16 km) away, claimed afterwards that they heard profanities couched in an Australian accent echoing among the mountaintops for several minutes.

However, this happened later, and the problems they encountered during the first three seasons were usually the result of over-eagerness to take every job that became available. As an example, in 1958 a Calgary-based mining exploration consulting company called J. C. Sproule & Associates (irreverently known among helicopter crews as Jesus Christ and Associates), through a series of mishaps, ran out of helicopters up the Mackenzie River in the Norman Wells area. Sproule's crew were stranded in the bush for six weeks without a machine. Bob Taylor finally went up at the end of the season which, as he said, "spread the company a little thin, stretching our logistical supply line all the way from Pat Bay to Norman Wells."

Sproule had a reputation for running very economical operations, Bob recalled: "There were six of us in camp and he [Sproule] would come off the plane with six steaks and six beer. That was a big, hearty surprise treat for the boys. It was at least a change from the eternal Chuckwagon stew!" (A remarkable number of the people we interviewed spoke disparagingly of Chuckwagon Dinners. Presumably the producers dealt only with exploration company purchasing agents, people who never had to go out in the bush and sample the product.)

By 1960 the operation had settled down to five machines, supplemented by as many as four leased from Aero Copters. Most work by this time was within the province.

The Stikine River was being extensively explored – there were several large drilling projects in progress – and both provincial forestry and topographical survey contracts had been renewed. Bob Taylor became involved with experimental projects with the Forest Service:

> They wanted to be able to drop food packages to people who were too far from a lake or a clearing to be reached [Bob recalled]. These packages weighed about 30 or 40 pounds [15-20 kg] each. We tried free-dropping them, but that didn't work very well – too many of them ended up smashed on the rocks. So we came up with the idea of attaching small parachutes to the packages, hooking them on to a bomb rack – we must have cornered the market on surplus Canso bomb racks – and tying the static lines to the cargo racks. Then all we had to do was to punch off the packages and the parachute would be yanked open automatically by the static cord. Trouble was, we had the packages on a short cord, hooked onto the bomb rack, and we hadn't figured out that the airstream would twist the loads round and round.
>
> The first time I tried it, I had two of these packages hooked up. They obviously began to twist as we took off. Just before I reached the drop area, one of the 'chutes opened by itself – the package had twisted until the static line tightened up enough to open it. So there we were all of a sudden, with the nose going down and when I looked back the 'chute was open and oscillating right by the tail rotor. I was afraid to punch it off in case it wrapped itself around the tail rotor – and I didn't want to let down too quickly for the same reason. The problem was to keep the nose up; I had the cyclic [stick] right back in my stomach. I tell you, there were two guys compressing their backs against the firewall. In the end, I sort of sideslipped and punched off both packages. The guys on the ground thought it was a big joke. They thanked me for waiting, until I got close enough for them to retrieve the packages.

This was one of several experiments conducted by the Forest Service with helicopters. Another successful one in which Bob Taylor was involved was the use of a camera boom for stereoscopic aerial photography. At first the boom was attached under the frame of the helicopter; later it was slung from the cargo hook. The boom, which was about 12 feet (3.6 m) long, had two cameras mounted on it, one at either end. The cameras were operated electrically, by remote control from inside the helicopter. This made it possible to cover the ground with overlapping photographs which, in turn, when fed into a stereoscopic plotting machine, made it possible to measure the heights of the trees very accurately. Using colour photography, the species and ages of the trees could also be estimated with some precision. In the end, this system permitted forest inventory to be conducted without the need for a classifier in the helicopter – a saving, both in time and manpower.

Some years later Bruce Payne took off with a new test version of the camera boom slung on the cargo hook. He was flying a turbine-powered Alouette, a much faster machine than the Bell, and the purpose was to test the boom for high-altitude photography at 115 knots (210 km/h). As it turned out, this was fast enough to induce the camera to fly on its own. The boom

To get your endorsement in those days, all that happened was that Alf [Stringer] sent you home with ten questions and the manual, and if you could find the answers in the manual and he was satisfied with your work on the floor, the DOT [Department of Transport] endorsed your M licence for rotary wing.

Eric Cowden, Engineer

Peter Moore opened the base at Prince Rupert. He'd been there for about a year, but they weren't doing too well and I was sent up, and we got into the old hangar, the old Air Force hangar at Seal Cove. I spent three years there. We did well. The three years I was there we did 2,500 hours. One year we did just under a thousand hours. I guess you could say the base idea paid off.

Bruce Payne, Pilot

came up and thumped the bottom of the helicopter; fortunately no serious damage was done and the boom was then modified so that it could be clamped to the bottom of the helicopter.

Another, more conventional job Bob remembered with a faint shudder was the installation of a Pacific Great Eastern Railway (later B.C. Rail) microwave tower:

> If you look up the side of Lillooet Mountain, you'll see that it was pretty hairy. The sidehill is just about vertical. They blasted a little 14-foot by 14-foot [4-m × 4-m] pad off the top of a shoulder and it was a straight drop 3000 feet [900 m] down to the valley floor – I mean straight down. They lost a couple of packhorses getting up there to blast the pad.
>
> When you landed there, you had to be careful about stepping out because there was practically no place to stand – and the winds you'd get were unreal; you'd toss a piece of plywood or something over the side and it would come spitting back up at you. The pad was on the northeast side, so you nearly always had subsiding air. You hardly ever got an updraft to help you get the tower steel up there; you had to dog it all the way, up. And sometimes, when you got there, you'd get all set up on your approach and then at the last minute suddenly lose 200 feet [60 m] and have to start all over again. We went through eight pilots on that job. Some of them would just take one look at it and catch the next PGE train out.

While these things were happening, a meeting that had taken place shortly after Ted Henson's death was to have a good deal of influence on the future of Vancouver Island Helicopters. When the life insurance policy on Ted, kept alive by Fred McGregor, was paid, the premium was used to purchase a new helicopter in 1958 – a Bell 47 G2, registered CF-KNC. By the time the machine arrived in Victoria, however, it was so late in the season that Lynn could not place it on a contract. Rather than leave it sitting in the hangar for the rest of the summer, Lynn offered it on lease to Okanagan. She and Bob Taylor went across to Vancouver to discuss the details with Alf Stringer, Okanagan's operations manager.

> It was a very business-like meeting [Lynn recalls]. We made the necessary arrangements for the lease and then left. But later, in 1959, there was a celebration of 50 years of flight in Canada, and I received an invitation to attend the banquet at a Vancouver hotel with Alf Stringer. I was very flattered that this important man from Okanagan would invite me; and it was an exciting evening: Igor Sikorsky was there as one of the guest speakers, so I met him, too. Anyway, the eventual result of that meeting was that Alf and I were married the next year and I went to live in Vancouver for the next three years.

Six months after her second marriage, and just before the move to Vancouver, Lynn suffered another tragedy. Her 10-year-old son was struck down and killed by a truck in Gordon Head. However, life had to go on and she moved to a large house which she and Alf had had built in the Oakridge area of Vancouver.

For the next three years, Lynn Stringer ran VIH largely by phone. Then, because Carl Agar's health had broken down and he had retired from Okanagan, and because Alf was finding himself increasingly at odds with management, Alf decided to resign. "It was," as Lynn observes, "an awful wrench for him. He had built up this little company until it was one of the biggest helicopter companies in the world. So it was quite a letdown for him to come and operate five helicopters after being operations manager for Okanagan."

Alf Stringer became president of VIH and Lynn, as she puts it, "stepped out of the picture and stayed home and had children."

Helicopters, like this VIH 206 flying past Victoria's Johnson Street Bridge, have in recent years been able to function as very effective camera platforms. This is a 360 degree Imax camera mount in use for a Disney movie in the early 1980s.

VIH COLLECTION

CHAPTER THREE

A Growing Business

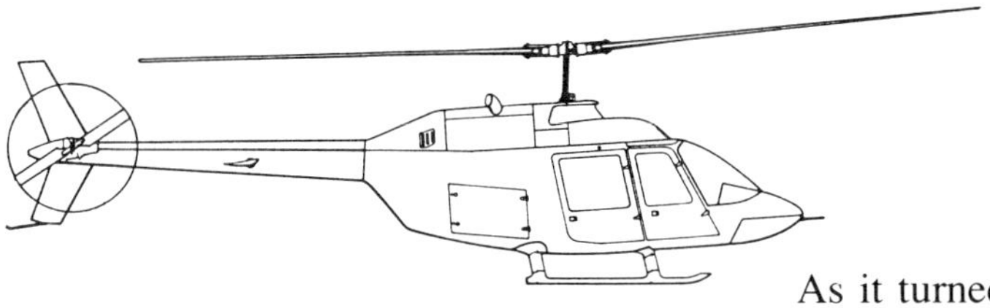

As it turned out, Alf Stringer's take-over period with VIH tended to be an anxious one. The early sixties were recession years and profits were scant. The company owned five machines, but contracts were hard to come by. Governments were practising restraint and so were the resource exploration industries. By 1964, just before Alf took over, there was some improvement. The company was able to renew its contract with the provincial Forest Service to conduct an inventory of timber throughout the province – a contract that endured for several years. There was renewed interest in the Stikine River, south of Telegraph Creek, where considerable mineralization had been located and the area was now being drilled and assessed. Some drilling was also done as a preliminary to damming the Stikine for hydro-electrical power to serve potential mines.

Another approach was to send one or more machines to a mining hot spot, Smithers, for example, or Burns Lake or Stewart; but long ferry flights did not make this concept too profitable – particularly since, as usually happened, the competition put machines in these hot spots as well. There were some exceptions. In 1967, pilot Dan Dunn, with Don Sadler as engineer, was sent up to Alice Arm, north of Prince Rupert. Molybdenum had been located across the inlet from Alice Arm, and this had set off a an exploration boom. Some five drill camps were spread in a semi-circle around the base, all less than 30 miles (48 km) away. Supplying these camps provided enough work to keep one machine busy; but there was also an almost continuous stream of geologists and mining engineers coming in to examine old staked properties in the area; so a second machine, crewed by Peter Corley-Smith and Don Borden was sent up, and both machines logged more than 500 hours apiece during the four-month season.

In fact, though, the company had started to look hard at the concept of year-round fixed bases as early as 1963 – in addition, of course, to the main base at Sidney. The first attempt was at Campbell River, but the application was turned down by the Air Transport Board because another company had applied earlier. Next, Prince Rupert was selected, and this time success came, but only after one or two lean years. Pilot Peter Moore went up and set up shop in the old Air Force hangar at Seal Cove. Business was slow, even though, as Alf pointed out, "The potential was certainly there, but there were a few operational problems that needed to get ironed out. We just had to go through a learning curve."

By the time Peter Moore left for new pastures and Bruce Payne took over in 1965, things began to look up: the economic recession was over and the base began to put in several years that yielded almost 1,000 revenue hours a year. Meanwhile, Alf Stringer gradually built up the fleet to six machines and an additional three leased – he was wary of too rapid growth because he had seen the problems, both in Okanagan Helicopters and with others like Spartan Air Services. This meant not only stability but quality: his helicopters were so meticulously maintained that they were always a source of pride to their crews – and customers took note.

The company's second year-round base was established at Port Hardy in 1968 by Peter Corley-Smith and Don Borden. Once again, the timing was propitious: the discovery of a substantial deposit of copper had set off a boom of prospecting while Utah Mines prepared to open a large, open-pit mine. For the first three weeks, things were depressingly slow – Okanagan had a Hiller 12E based there, flown by Bill Foote. Discouraged by the lack of business, he left and, shortly thereafter, customers appeared in droves. There was more than enough work for one machine, but that was a busy year throughout the province, and the company couldn't send another one in. Port Hardy was followed the next year by a fixed base at Stewart. There were good years and bad years, but on the whole the base concept worked and it had the advantage of being far more agreeable for the crews, who were now able to have their families with them year-round – besides which there was far more variety than usually found in a fixed contract.

As Alf explains, "From the time I arrived over here, we were gradually fading from fixed contract to base work – trying to get something that would give us year-round activity; that was the big thought behind the base concept." In the long run, the fixed base, with hangar and full services "made the

I went back to university the year I joined VIH; flying in the summer, school in the winter. In all the 12 years I worked for them, I never had a written contract with the company. I was paid what I thought I was worth, and when the summer came I never had to worry about a job. I could leave university and go straight back to work. Towards the end of the season, there was always a pilot to relieve me. In an industry that was notorious for its sudden shifts of policy, and abrupt lay-offs, I was remarkably lucky to enjoy such stability.

Peter Corley-Smith, Pilot

When we opened the base at Port Hardy, things were depressingly slow for the first three weeks. Then we met two of the stewardesses from the daily CPA sked in the coffee shop. They wanted a ride in the helicopter. Why not? Others followed in the next week or so; it was fun; they were attractive young women. Then it occurred to us to ask them to see if they could find any potential helicopter customers on incoming flights and steer them our way. They did, and we never looked back; we had more work than we could handle for the rest of the season.

Peter Corley-Smith, Pilot

The Bell 47 G2 got too expensive for training. The parts prices kept going up and we just couldn't make it run. We actually didn't train for a couple of years because the 47 got so expensive. And then we had a big problem recruiting pilots, not having confidence in individuals. [If we trained] we could have a better look before we hired them. That's when we got into the R22 and we had up to three and now we're down to one.

Barry Hewko, General Manager, VIH

BOB TURNER

General manager Barry Hewko beside a map with coloured pins indicating the location of the VIH's helicopter operations throughout the province.

difference between whether you made it or didn't make it in this business." This approach has prevailed to the present: the company now has nine fixed bases up and down the B.C. coast.

In addition to the changing base concept, new uses for the helicopter were still evolving. One of them, mentioned earlier, was the development of the camera boom for stereoscopic photography to be used in forest inventory work. Hugh Lyons of the Forest Service Valuation Branch was the first to propose this method. He recalls:

> The idea of the camera boom goes back to an article I was reading in *Photogrammetric Engineering Journal* by Eugene Avery. He was working in the southern states and had experimented with a hand-held boom with two cameras, and had succeeded in getting measurements by laying out plots on the ground; in other words, using ground control. I talked it over with Bill Hall, the head of Aerial Surveys. So Bill and I figured that if you could get the cameras far enough apart, and knew that the same distance would always separate centres, we felt that we could dispense with ground control.
>
> We then got hold of a couple of Fairchild F-24 cameras with five-inch film. Bob Taylor and I put our heads together with Ron Adams who, like Bob, worked for Vancouver Island Helicopters. Ron built the first camera boom to my specifications. We mounted it on the helicopter by clipping it directly to the cross tubes below the cockpit and to the nose and tail by means of straps. This was necessary because there was a long cantilever.
>
> There was an electrical firing mechanism; wires stretched from the cameras on the boom up through the door into the cockpit and a control box. The system worked very well; accuracy was excellent; but I was still looking for even better results. The cameras weren't fast enough and vibration was a problem.
>
> So we went to Linhof cameras to actually begin operation with the booms. Before this, all our efforts were purely experimental. There were still problems – the cameras wouldn't fire properly. If the rear camera fired too early or too late it would distort the base length. When things did work properly, it was possible to get accurate gross volume estimates of timber by species, and the results were comparable with ground sampling methods. We checked this by flying a section with the boom and then checking the same area on the ground. We could get the same results for roughly 50 per cent of the costs experienced using ground methods.

The camera boom was progressively improved by the Forest Service. After the initial experiments with a fixed boom, which had resulted in vibration problems, a new approach was tried. A boom was produced that was attached by two lanyards to a longitudinal bar clipped to the cargo hook and suspended beneath the helicopter. This helped resolve the vibration problem, but the boom required a certain and constant airspeed before it would fly without oscillating. Eventually, in 1976, another boom was

Scenes such as this draw many to helicopter bush flying. This is a Bell 47 G3B-1 of Northern Mountain Helicopters at Thutade Lake.

JOHN KING PHOTO

This is an early, possibly the first, ▷ version of what became known in B.C. at least as the "Monsoon Bucket." Jim Grady of Okanagan Helicopters and Henry Stevenson, owner of a Nelson machine shop, perfected the device in 1962 after years of work. The pilot dunked the converted 45-gallon drum in water, flew it to a fire and then opened a trap door in the drum, releasing the water onto the blaze.

B.C. MINISTRY OF FORESTS

Later versions of the camera boom were designed to attach directly to the helicopter to avoid the problems experienced with earlier designs.

B.C. FOREST SERVICE

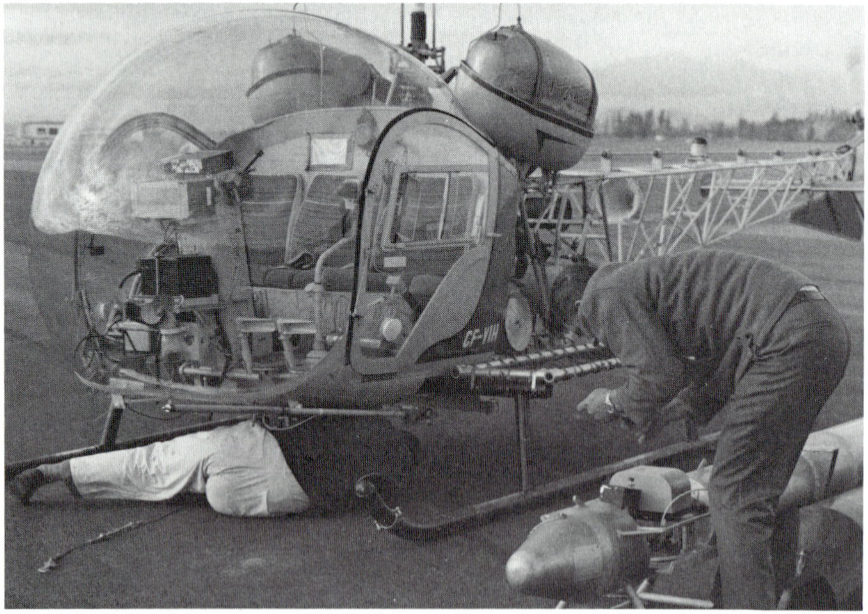

The bucket, shown here below a VIH 204B, was a successor to the monsoon bucket. It was used in the same way as a monsoon bucket but was more easily stored when not in use, had a greater capacity and allowed for adjustment of the amount of water carried.

VIH COLLECTION

Hank Sands, Base Manager at Stewart in the early 1970s, standing beside a Bell 47 G3B-1. VIH COLLECTION

Pilot Cliff Schwindt (left) and base manager Irv Tingstad, standing in front of the tail boom of a Bell 47 J2 at the Seal Cove Base at Prince Rupert. VIH COLLECTION

Drip torches, among other imaginative devices such as gasoline-filled ping pong balls were used to start slash fires, to destroy the debris of logging. For environmental reasons, slash burning is no longer a common practice.

P. ROBIN PHOTO, B.C. FOREST SERVICE

Many of the provinces' communications towers, vital in the rugged British Columbia mountains, have been installed using helicopters like this 206 working on a communications repeater tower.

B.C. MINISTRY OF FORESTS

This Bell 47 G3B-1, shown in 1974, is hoisting high-value cedar blocks, later to be cut into roofing shakes. While business was good in one sense, it was often difficult for helicopter companies to get paid for this work.

B.C. FOREST SERVICE

VIH established an important base at Seal Cove. From this facility, it could serve the Prince Rupert region and the Queen Charlotte Islands before a base was established at Sandspit.

VIH COLLECTION

The Hughes 500D, favoured by some VIH customers over the Bell, was faster, could pull a heavier load, but was more expensive to maintain. Some pilots preferred the less crowded cabin of the Bell.

VIH COLLECTION

DOT inspector Ray McGowan (left), Bruce Payne and Ken Norie, just after Norie's check ride for his helicopter endorsement on his 17th birthday.

NORIE COLLECTION

In the deep snow. Trevor Deeley's 206, CF-VIH, on lease at the time, is in the foreground.

VIH COLLECTION

Operating out of VIH's Stewart base, this 205A-1 is hauling a drill on the long line during the Red Mountain operation of the early 1990s. Red Mountain is a few miles north of Stewart.

VIH COLLECTION

Trevor Deeley's 206 CF-VIH, parked in front of his waterfront home in Sidney.

TREVOR DEELEY

There is little spare room in VIH's Pat Bay hangar before the heaviest operating season of the year, the summer, arrives. Aircraft, the ubiquitous 206s, undergo major inspections and maintenance.

BOB TURNER

Engineer Dave Holden at work.

BOB TURNER

Race Rocks lighthouse, which has guarded the approaches to Victoria for over a century, has recently been a popular site for both movies and television commercials. Nearly all B.C. lighthouses have helicopter landing pads.

VIH COLLECTION

Below: Lycoming engine in a Bell 47 G2.

DAVE PARKER

Below right: Soloy turbine conversion in a G3B frame.

DAVE PARKER

At 9:30 a.m. on March 14, 1991, fire broke out at Joan's Flower Shop on Prince Rupert's Third Avenue. Firefighters had difficulty reaching the site because of road construction, so VIH, based at Seal Cove, was called in to water bomb the blaze. VIH responded with a 206 and a 205, the latter flown by Brian Mycroft. Mycroft recalls that, guided by a firefighter spotting on the ground, they were able to make drops at one-minute intervals, the only problem being the toxic smoke, which meant they had to hold their breath while dropping. The fire was contained with little damage to adjacent buildings within an hour.

INNERVISION PHOTO, VIH COLLECTION

The 204B was, with civilian avionics, larger main rotor blades and a plush interior, the commercial version of the military UH-1B. This VIH machine, shown at Stewart, was one of only 192 built. The 205A-1 succeeded the 204B in production.

VIH COLLECTION

◁ The LongRanger was a logical development of the standard five-passenger 206, and was designed to fit midway between the 206 and 15-passenger 205A-1 in the Bell lineup. The original LongRanger was introduced in September, 1973, and the improved 206L-3 during 1981.

VIH COLLECTION

Frank Norie and his "daily driver," a Bell 47 G2. DAVE PARKER

On July 27, 1992, a major brush fire broke out on Victoria's Mount Douglas. Aerial attack using both fixed-wing tankers and helicopters with buckets proved necessary. It was very hot, frustrating work for these Saanich firefighters on the side hill. JOHN MCKAY PHOTO, TIMES-COLONIST

designed specifically for use on the Bell 206. The system is attached by means of a semi-rigid mount to the helicopter in a matter of minutes and enables inventories to be carried out as an ongoing process over the six forest regions in the province.

As well, the use of other helicopter-mounted devices has substantially simplified the work of the Forest Service and, in one case, replaced a somewhat surprising tool of foresters – the rifle. For years, foresters have been searching for what are called plus trees – trees that are as close to the ideal as possible – for each of the various species found in the province. By selective propagation they hope to produce a super-tree for each type that can be "cloned" to produce higher quality timber. This genetic information is acquired from cuttings, or scions, taken from the tallest, straightest, generally closest to perfect tree in a stand. Foresters on the ground would search out such trees and, using a rifle, shoot down a cutting. With helicopters, cuttings could be removed directly from tree tops by means of a harnessed crewman hanging out of the cabin. Before long, shears were developed to clip cuttings from trees, a device that worked off the helicopter's hydraulic system. Later still, baskets were developed to harvest cones so that trees no longer had to be climbed by pickers or felled so that cones could be harvested on the ground.

Yet other useful functions were developed in the forests. Almost from the beginning, people appreciated the value of helicopters in firefighting – particularly in transporting crews and their equipment to deal with a fire as soon as it has been spotted and before it becomes unmanageable. Before long, more varied tasks were tackled. Helicopters were ideal for observation – those in charge could monitor the progress of fires and plan strategy accordingly. They could decide in advance where to place crews and equipment and where to create fire breaks. If necessary, the helicopter could be used to promptly remove a threatened crew from the path of a fire that had changed direction. The same could be done in the case of accident and injury.

The first use of a helicopter as a tanker in British Columbia occurred in 1958 when Okanagan Helicopters converted a Sikorsky S-58. The cabin was fitted with a 225-gallon (1 025-l) tank and appropriate plumbing and discharge gates were cut in the belly of the aircraft. The S-58 was used to extinguish a fire near the headwaters of the Squamish River. When asked about the conversion, Alf Stringer commented that it created a sig-

DAVE PARKER

We used to take the door off and the Forestry guy would lean out as you hovered by a tree and try to cut a small branch or grab the cones off it. The next thing we tried was putting him in a bucket on a 50-foot [15-m] line clipped to the cargo hook, and then dangling him slowly up the side of the tree. The guys thought this was great, but the Workman's Compensation Board took a dim view, so we had to chuck that one.

John King, Pilot

nificant problem. With its conversion, the S-58 ceased to be a general purpose machine and became exclusively a tanker. Operationally, this was impractical.

The solution, like most good ideas, seemed obvious enough in retrospect. Jim Grady, base manager for Okanagan Helicopters in Nelson, and Henry Stevenson who ran a machine shop in the city, are generally credited, in British Columbia at least, with the invention in the early 1960s of the "monsoon bucket." In its earliest form, it was simple: a 45-gallon (205-l) drum filled with water, with a basketball plugging the hole in the bottom. Hovering over the fire with the bucket suspended from the cargo hook, Grady would lean out of the door and pull on a line attached to the ball to uncork the drum; whereupon the water cascaded down onto the fire, extinguishing it or, at least, retarding it. Before long the prototype was improved: a small hatch resembling a trapdoor replaced the basketball. The trap was activated by an electrical solenoid, controlled by a pushbutton on the control column to release the water. More recent versions bear little resemblance to the original; they are of lightweight plastic and can be rolled up and stored, and have various capacities; in some the volume can be altered by the removal of plugs in the sides of the buckets.

BILL BUHR

Bill Buhr, one of more than 20 pilots flying the ambulance service, beside a Bell 222-UT.

The monsoon bucket has several advantages over the extensive S-58 conversion. Any general utility helicopter can use one since all that is required is a cargo hook and an electrical connection. The bucket can be quickly attached or detached. It can be filled by simply hovering over water and dunking it. Not having to land decreases turn-around time between loads, and the bucket, when not in use for firefighting, can be used to transport supplies and equipment without landing. The primary use of a monsoon bucket is extinguishing lightning strikes before the fire has had time to start moving, or to attack small flare-ups and "hotspots" after the main fire has been extinguished.

A disadvantage of the monsoon bucket is that it releases water slowly as compared to a fixed-wing tanker. This increases the time the helicopter must remain over a blaze, which in turn may result in the fire being fanned by the downwash and embers being blown about to start new fires.

At the other end of the spectrum, helicopters are efficient tools for setting fires. In 1938 a serious fire broke out on Vancouver Island. The outbreak advanced rapidly in accumulated logging debris not far from Campbell River. It spread out of control until it was threatening Courtenay. Fortunately the town survived and the danger posed by slash (the debris of logging) resulted in constructive action. The British Columbia Forest Act had a section added (Section 116) to ensure that the Forest Service could require the destruction of debris – the most obvious method was to burn it. This practice would help to deprive the fire of fuel. Slash burning, which itself became controversial and has since been largely curtailed in favour of less ecologically damaging techniques, was normally carried out with hand-held Panoguns, devices similar to flamethrowers and fuelled with diesel oil. This was a time-consuming process and often required large crews.

In 1973 John Muraro of the Canadian Forest Service at Prince George developed what has been termed a drip torch: a 45-gallon (205-l) oil drum bolted to a metal frame and connected to a long spout that dispenses a 60/40 mixture of diesel oil and gasoline. This apparatus is suspended below a helicopter and the fuel is ignited electrically by a solenoid activated by the pilot. The flaming liquid then drops to the ground, setting fire to the slash. Using the drip torch, slash burning can be more selective; it isn't necessary to fire the whole setting as would often be the case with ground-based methods, a process that could result in an uncontrolled fire and even soil damage. With the drip-torch operational, the pilot is directed by radio, sometimes setting a number of fires simultaneously. Using ground methods, it would have taken perhaps ten men to achieve the same result. There was also some element of risk to the ground crew and there is virtually none for the helicopter pilot.

Yet another task for the helicopter that promised great potential was in the realm of power transmission lines – both with construction and with maintenance. The first venture began as early as 1958, when Okanagan Helicopters was awarded a contract by the Southern California Edison Company to construct a power transmission line through Soledad Canyon, northeast of Los Angeles. Concrete bases were poured by a Sikorsky S-58, then pre-rigged wooden poles were dropped into the prepared sockets in the concrete. The operation was a success and was widely reported in North America. Moreover, it led to the awarding of many similar contracts in Canada, particularly in British Columbia, with its difficult terrain.

BOB TURNER

Before he settled down to fly a desk, Operations Manager Charlie Mooney had his exciting moments as an operational pilot.

One of the most hair-raising experiences I had was when I was with FIFT [Forest Industries Flying Tankers] flying a JetRanger. We were starting a prescribed slash burn, using the old "Ping-Pong" system – just starting to try and get a fire going up near Cassidy. It had nothing to do with that apparatus but I had an in-flight explosion and fire. I was about 120 feet [37 m] above a steep sidehill, and there was only one little spot, a logging spur to land on. While I was heading down the hill towards it, my engine quit and we did a successful autorotation. We got down, upright and in one piece. I had one passenger, Paul, who was operating the ignition machine and he was in the rear seat – both of us on the right. His door was off and my door was off. It was slopey ground where we touched down. I touched the right skid first and we bounced about a foot into the air. I had my head out of the door because of the smoke and out of the corner of my eye I saw Paul diving out of the machine before we had settled. I remember thinking "yeah, you chicken, you didn't stick around for the whole flight."

Charlie baled out shortly thereafter and 15 minutes later "there was nothing but a pile of ashes left."

Charlie Mooney, Operations Manager

Bruce Payne was involved in the early stages of this development. Once the survey for a line had been completed and the tower sites located, landing pads were built at each site. Ground crews would then be flown in to construct the forms for the concrete footings. Then the concrete, already mixed at a central point, would be flown in. Bruce describes the method:

> Most of the buckets at that time were 45-gallon [205-l] drums. We'd just put a ring on the bottom of them. Then you'd lay the drum on the rim of the form, a guy on the ground would transfer your lanyard clip onto this ring and you'd lift up and dump it for them. Later on we got the self-dumping buckets. You'd just hover over the form, a guy on the ground would pull a lever and it would dump itself.

Once the forms were poured, the helicopter would start to sling in the steel for the tower. Sometimes, with the smaller helicopters, they would actually hang the tower steel length by length, while the ground crew bolted them in place. Later, with the advent of more powerful turbine helicopters, the whole tower would be swung onto a prepared base. This was moving into the flying crane concept and offered an exponential leap in productivity on many other construction projects besides transmission towers.

To continue where Bruce Payne left off, the next stage in power-line construction would be the installation of the cables. Here, once again, the helicopter could speed things up enormously. The process began with the stringing of what is called the sock line – a thin steel cable , perhaps a quarter of an inch (11 mm) in diameter, which would later be used to haul the actual power cables into position.

The sock line would be on a large drum, mounted on the back of a flatbed truck. The drum had a brake on it, operated manually, to allow the line to be paid out at a steady tension. A metal ball weighing 150 to 200 pounds (70-90 kg) would be slung below the helicopter. "The ball hung about 20 feet (6 m) below the helicopter and they'd hook the line onto that. Then I'd just move out and pull it off the ground. I've pulled out as much as 42,000 feet with a Bell 47. About seven or eight miles (11-13 km), I guess. It would take up to five minutes to get it moving."

On at least one occasion it took longer than that. After pulling maximum climb power for some ten minutes, Bruce finally gave up, dropped the ball and flew back to the drum truck to find out what was wrong. When he got there, he found the operator sound asleep, with the brake firmly on.

When it came to the sock line moving, the pilot would pass each tower and then back up to give ground crews enough slack to run the line through a "traveller" or pulley, which would then be pulled up and secured to the cross arms on top of the tower. Once the length for that particular section had been run out, another truck with a winch would hook on to the sock line and pull the actual power cable, which was attached to the tail end of it, through the travellers.

> It wasn't a bad job [according to Bruce], but some of the guys couldn't take it. They could pull sock line over the flat ground, or in the not-so-hilly stuff – but when they got over one of those big valleys, with a thousand feet [300 m] of nothing beneath them, and they're sitting there at, oh, maybe three or four miles an hour [5-6 km/h] for 20 minutes [to] 30 minutes, pulling it across, they'd get vertigo pretty bad and have a real problem.

Meanwhile, technology brought other changes. The piston engine was giving way to the turbine. The Bell 47 series had progressed through some 18 piston-engined model changes since the original, open-cockpit B3, the latest being the turbocharged G3B-2, which had taken much of the terror out of mountain flying. When Bell introduced its first turbine helicopter, the 206 JetRanger, Alf Stringer wisely held back for a while. Instead, in 1972, he bought an S.A. Alouette III, a turbine-powered helicopter that had been in service for some years and had proved itself. Alf remembers it as an excellent machine that just didn't fit in at the time.

> It had great performance; it was beautifully stable in the hover and we had hoped there would be a demand for it – hydro and those sorts of jobs – but it didn't work out. It used huge quantities of fuel, which made it very costly to operate. It had a great big engine, which was de-rated, and this gave it a tremendous altitude performance. I guess it was just too early for where we were based.

The reason for holding back on the Bell 206 was that Alf, through his contacts in the American Helicopter Association, heard that the early turbine's engine was giving all sorts of problems, and VIH was too small a company to be able to afford a sort of trial and error exercise. "I think one of the main problems was that Bell and other people demonstrating the machine would come in and, as soon as the collective was down, shut off the engine. Same thing on take-off – start up and take off right away. They weren't giving the engine any warm-up or cool-down period, so they were cooking the engine. Once this had been appreciated, and also with the introduction of the

DAVE PARKER

The plumbing on a Bell 47 G3B-2's turbocharger.

My wife wouldn't let me land on the front lawn [of his Sidney waterfront property], so I landed down on the beach. One day, when the tide was coming in, I went to move it but the battery was flat. A quick call to the VIH hangar brought Eric Cowden racing over with another battery. He was standing up to his knees in water as he changed them – we got it off just in time.

Trevor Deeley, Vice-president, VIH

The first time I used a JetRanger, in 1969, I think it was, they were logging 20% in-flight engine failures – it was horrendous. But once they'd ironed out the engine problems on the C18 engine, and then brought out the C20 . . . it's been excellent. Statistically, the JetRanger is now the most reliable single-engine aircraft in the world.

Bob Reimer, Chief Pilot

Bob Reimer.

improved C-20 engine, we saw that it was a good machine: the airframe gave us no problems at all – so we started to get into them."

The first two Bell turbines, 206Bs, or JetRangers, were purchased in 1973. By 1975, the company owned five 206s; by 1978, this had increased to six 206Bs, two 206L-1s, or LongRangers, and one 47 G4. Another venture during this period was hauling blocks for cedar shakes out of the bush. To take this on, the company invested in two conversions with the Bell 47 G3B airframes to the Soloy kit – a kit that replaced the Lycoming piston engine with a lighter but more powerful turbine. The machines worked well, Alf Stringer noted: "They were very successful because you could sit up there and hover all day: the engine was de-rated enough that you were just cruising." Unfortunately, though, the business turned out to be too speculative. "We only had them for a couple of years, then we sold one down in the States, and the other to Alpine Helicopters. The problem was that the shake business got so competitive that we had trouble getting paid. You could work forever for these people and not get any money out of them because they didn't have any – they were working on the next two blocks down the road."

The company's fleet expanded steadily. In 1984, it moved into the medium lift field by purchasing a Bell 204B, with a working payload of 3,200 pounds (1 415 kg). Four years later the 204B was replaced by the more powerful 205A, with a payload of 3,600 pounds (1 635 kg). In this period, too, VIH began to use the Hughes 500D (now called the McDonnell Douglas). They did so with some reluctance because they would have preferred to standardize the fleet with Bell helicopters; but in Chief Pilot Bob Reimer's words:

> When Hughes came along with the 500C, they put it out cheap, and it would do what the JetRanger would do and even a bit more. It was a $150 to $200 an hour cheaper and the mining community just ate it up. It was smaller and would go into a smaller clearing. Then they brought out the D, and they thought that was the best thing since sliced bread because it could outperform the JetRanger considerably – it could pull about 300 pounds [136 kG] more and it was faster than the JetRanger. They were definitely better for moving drills – 300 pounds better. The only problem is that they require a lot more maintenance than the Bell.

As the fixed bases settled down for the long haul, they became the stepping stones for progress in the company. In Prince Rupert, Bruce Payne returned to Pat Bay to become chief pilot, and he was replaced by Irv Tingstad, who remained the base manager at Prince Rupert for many years before leaving the company to involve himself in a logging business and Ken Norie took his place. In 1972, Don Honeyman took over from Ron Jeffries, who had run the base at Port Hardy for two years, and remained there until 1980, when Payne left to join MacMillan Bloedel and Don replaced him as chief pilot. Honeyman remained in the position until he left the company in 1985, and the check pilot, Bob Reimer, was appointed. As Reimer recalls; "Along with the chief engineer, Barry [Hewko] – there was really only the two of us and we were operations; the two of us looking after the whole thing. Then they hired a chief engineer [Larry Huff] which certainly helped because that off-loaded Barry a lot."

Reimer's functions changed significantly over this period, as they would in any company that is expanding.

> The first three years I was chief pilot, I did a lot of line flying away from the main base, and a lot of the base flying [Victoria], along with John King and also all the *ab initio* training, the recurrent training and all the check rides. We did the training on 47 G2s, and then, the year after I quit training, we got the Robinson 22s. Going back about six years, from that point on I did less and less flying so that now I'm only doing about 100-150 hours a year. When I started, we had 12 helicopters and I think there were 10 pilots on staff; by December 1984, I had 65 pilots to look after – about 48 full-time and the rest contract – a lot of administration. For example, with the medevac operation, there's three machines and something like 23 or 25 pilots. To begin with it was a bit of a logistics nightmare, but now it's settled down and pretty well running itself.

During this period, too, there were considerable changes at the executive level. In the June 1970, Frank Norie, a successful businessman and partner of Norie Bros. Logging, had bought a Bell 47 G2 helicopter and learned to fly it at VIH. He used it to visit his logging operations up Island and, not surprisingly, all his maintenance was done by VIH. In 1975, Bill Boeing Jr., one of the founding shareholders in the company, decided to reduce his holdings in a number of diverse companies, and Frank Norie bought his VIH shares. This, naturally, made Frank a director of VIH. Then, in 1985, when Alf Stringer experienced heart problems resulting in a bypass operation and decided "that I wasn't going to try and run the thing full-time any more, we got Barry Hewko in as general manager." In October that year, Stringer, and his wife Lynn, who were major shareholders, were approached by a businessman from the United States with an offer to purchase their shares.

BOB TURNER

Engineers Ian MacDougal (kneeling) and Craig Brown at work on the Robinson R22, C-FHSC.

I had my fixed-wing pilot's licence and decided I'd like to go in for a helicopter. So I bought one with two partners [A Bell 47 G3B-2]. They both learned to fly in Ladner, while I spent the next few months in California, where I took training and got my American helicopter endorsement; in Canada, I only had to write the flight reg's for my licence.

When I came back, my partners didn't know this. I phoned them, complained about the condition our helicopter was in and arranged to meet them in Ladner. When I saw them arrive and go into the coffee shop, I fired up the helicopter, moved over and hovered right in front of the coffee shop window where they were sitting. They made a fuss about that one.

Trevor Deeley, Vice-president, VIH

The word got around, and Trevor Deeley, after consulting Alf Stringer and Barry Hewko, decided to involve himself with the company. Deeley, who owned the Harley-Davidson motorcycle dealership for Canada, had bought a Bell 47 helicopter in 1980, learned to fly it and was having it maintained by VIH. In 1984, he moved up to a JetRanger and he was enthusiastic about the business. As Barry Hewko describes developments:

> When this American investor became interested, we were a little nervous about what would happen; so I talked to Trevor about it and he said he would be interested in investing in the company. So we put a package together, myself, Trevor and Frank Norie. Frank took on more of a shareholding, Trevor came in and so did I, and we bought out Alf and Lynn – I guess it was in the fall of 1985.

Consequently, Frank Norie became president and Trevor became vice-president. As Hewko points out, Deeley was with the company "just a few years, but he was a big asset because of his financial powers. He gave us credibility with the bank and stability, particularly at a time when it was tough going – flying was down and expenses were high and he helped us financially."

Aircraft were frequently leased to meet sudden increases in business. The brothers Frank and Henry Norie, for example, leased Bell 206s, as did Trevor Deeley. However, by its 40th anniversary year, the company operated 26 Bell 206Bs, two Bell 206Ls, two Bell 205As, one Robinson R22 (for training), five Hughes 500Ds, one Bell 222A, three Bell 222-UTs, and its associated company, VIH Logging Ltd., owned a Sikorsky S-61Ns. The 222s are used for ambulance service and the S-61s for logging. These aspects of the business will be dealt with in the next chapters.

VIH COLLECTION

Chief instructor Uisdean Ross in a Robinson R22. From the start, VIH trained pilots, for many years on the Bell 47 G2, more recently on the R22s.

CHAPTER FOUR

Ambulance Service: A New Dimension

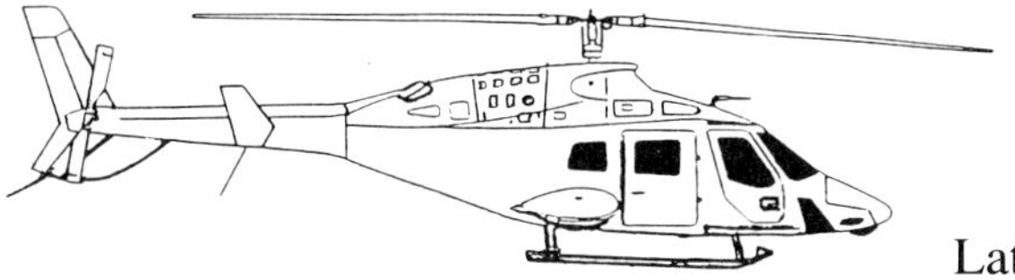

◁ The first official B.C. Ambulance Service helicopter, a Bell 212, is shown approaching the Prince Rupert Regional Hospital. VIH was not entirely satisfied with this machine and within three years changed to the 222. There is nothing in the livery to indicate this is a dedicated ambulance machine. It served the Prince Rupert area and Queen Charlottes.

BRIAN MYCROFT PHOTO

Late in 1988, Vancouver Island Helicopters won a contract that moved it into what was becoming an important new field in the industry: helicopter ambulance service. Over the years, VIH, in common with all helicopter operators, had frequently been called on to move people who were sick or injured from normally inaccessible places to medical facilities. However, during the last decade people had begun to equip helicopters solely as air ambulances, and put them on constant standby with both medical and flying crews to serve a specific area. The term "medevac" has come into the language and, recently, ambulance people have begun to refer to "airevac" to distinguish between air- and ground-based ambulances.

Since its inception in 1974, the British Columbia Ambulance Service has been extending medical service to all areas of the province: by road where possible; by air or sea when required. In 1977, the first air ambulances were introduced; they were Government Air Service planes supplemented by chartered commercial fixed-wing aircraft and helicopters. The government hesitated to expand into dedicated helicopter ambulance operations because the costs seemed formidable. Nevertheless, the advantages were so significant that, by the late 1980s, it decided to do so, beginning with an operation in a single, sparsely-populated region that, by its nature, presented problems for ground transport; and it was then that VIH made its successful bid to move into this field.

A year or two before this happened, the Prince Rupert Lions Club recognized the need for a better ambulance service in their region. They went further by contracting ALC Airlift Ltd. of Pitt Meadows to provide a dedicated helicopter ambulance service out of Prince Rupert – a service that also included the Queen Charlotte Islands. Although the contract was with the Lions Club, the Emergency Health Service Commission contributed funding, probably on an individual flight basis. ALC used a German-designed MBB (Messerschmitt-Bölkow-Blohm) 105 helicopter, a type frequently used for this purpose in the United States. However, when the government asked for bids, VIH was successful and Airlift dropped out of the air ambulance business in British Columbia.

Having won the contract in November 1988, and facing a start-up date of January 1, 1989, VIH had to acquire an appropriate helicopter – and do so quickly. So the company purchased a twin-engined Bell 212 in England and had it flown in an air freighter to VIH's hangar by December 1, 1988. There followed a period of intense activity because the machine had to be totally overhauled and converted for its ambulance role. A crew of 20 engineers worked round the clock until December 31.

The deadline was met and the 212 arrived in Prince Rupert at 11 p.m. on New Year's Eve. The first call for its services came scarcely an hour and a half later.

This was a promising start, but there were problems with the 212. Firstly, it was very noisy – a big disadvantage because obviously the helicopter had to land close to hospitals, often in the centre of communities. Secondly, it was comparatively slow; and thirdly, an auxiliary tank had to be installed in the cabin to increase its range – a tank that occupied room needed for medical equipment. VIH began to look around for a more suitable machine, deciding in the end that the Bell 222 would solve most of the problems.

The 222 was Bell's first venture in nearly 30 years to produce an aircraft designed from the outset to serve commercial rather than military operators: the first was the almost legendary 47 series. The 222 began operating in the late 1970s, mainly for the offshore petroleum industry and other corporate users who needed a fast, well appointed helicopter. It also appeared to suit the ambulance role, so VIH found one, a 222A, available in Ontario and purchased it.

Although the "A" was a success in Prince Rupert, it was on wheels rather than the operationally more practical skids and was a little short on power. Early in 1991, the government began to think of increasing the number of helicopter ambulances to three: One in Prince Rupert, one in Vancouver and one in Victoria. Late in 1991, the government asked for bids. General Manager Barry Hewko describes the thinking in VIH:

We brought the 212 in from England. It arrived on the first of December and we had to become an IFR- [instrument flight regulations] certified company, on line with a medical-equipped helicopter by the first of January. I had 20 engineers on it for a month – you couldn't get near it. We got to Prince Rupert at 11 o'clock at night on December 31 and did our first medevac at 12:20 on New Year's Day.

Larry Huff, Chief Engineer

BOB TURNER

Chief engineer, Larry Huff at his desk.

RANDY L'HEUREUX

VIH took the 222-UT to the Detroit Airmedical conference.

> We went through the whole process of thinking it through. What could we get, what could we do, how could we accommodate them? The 222 is a very good machine, cheaper to operate than some of the others; but we felt that for what they wanted to do, they needed something with a little more capacity, better single-engine performance, longer range, and so the [more powerful] 222-UT [utility] came to mind.

This decision wasn't difficult, but implementing it was. The price of new UTs was beyond the company's means and there were, in Chief Engineer Larry Huff's words, "only a couple of 222-UTs for sale in the whole world at that time. One of these was in Taiwan, and that was in tough shape. We knew that when we bought it but the price was right and we decided to bring it over and we went right through the whole machine and basically just overhauled everything."

At this time, while the 1988 Prince Rupert contract was in effect, and the new contract was being considered, the Ambulance Service provided neither encouragement nor commitment to any operator; in other parts of the province, it was chartering helicopters on the same terms as before when emergencies arose. Program Manager Gordon Bates explains the helicopter situation in general:

> As everyone knows, helicopters are expensive; consequently, we tried not to use the helicopter when it wasn't appropriate. Unfortunately, this led to a bit of a snowball effect: if the volume of flying was low, the costs were very high. When the operators didn't get the volume, they'd raise the costs and we'd try to use it less. At the same time, there were some situations where we didn't have the choice – we had to use the helicopter. The Gulf Islands are a perfect example: it's almost automatic.

The Ambulance Service concluded that operating costs could be kept down if it committed itself to one or perhaps two companies. More or less simultaneously, it decided on a policy it called 2-2-2 (which had no connection with the Bell 222): two engines, two attendants, two patients. Naturally, this posed problems for VIH because no medical kit was available for the Bell 222 and it wasn't large enough – unlike such machines as the Sikorsky S-76 – to accommodate the medical equipment as well as two paramedics and two patients. Barry Hewko's solution was to stack the stretchers, placing one lower and ahead of the other. The only drawback to this idea was the difficulty the two paramedics would face having to lift the second stretcher into place in the confined space available. The solution would have to be some sort of lifting and swivelling device.

As the job unwound, we found that the 212 was too noisy. Every time it landed at the hospital, it was very annoying . . . because it was in a small town, and because it was in the centre of town, and so we wanted something quieter – faster, smoother and quieter. We settled for a Bell 222.

Barry Hewko, General Manager

Once VIH decided to use the 222 for air ambulance service, and purchased machines began arriving at Pat Bay, they had to be totally stripped, reconditioned, rebuilt and equipped for ambulance work. This 222-UT, intended as a back-up machine, is shown in November, 1994.

DAVE PARKER

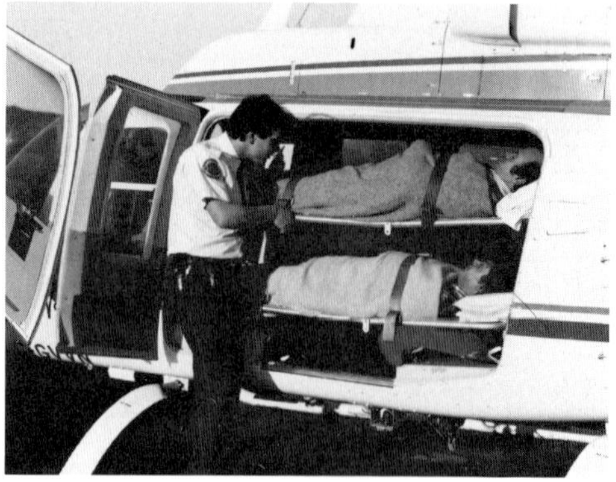

Before the creation of dedicated ambulance helicopters in British Columbia, helicopters in standard configuration, like this VIH Bell 206, were pressed into service on a charter basis. Though they did the job, the fully-equipped and ambulance-configured 222-UTs are much better suited to this demanding task.

BRIAN MYCROFT PHOTO

The 222A differed from the UT in that it lacked the more practical skid gear and was less powerful. These machines, though they carried some medical markings, were never painted in the full air ambulance colour scheme.

DAVE PARKER

Pilot Roman Bejnar shown shortly before a flight to Comox. Ambulance flights to points on Vancouver Island originate from VIH's Pat Bay hangar. Many helicopter ambulance flights are made to transfer patients between hospitals.

DAVE PARKER

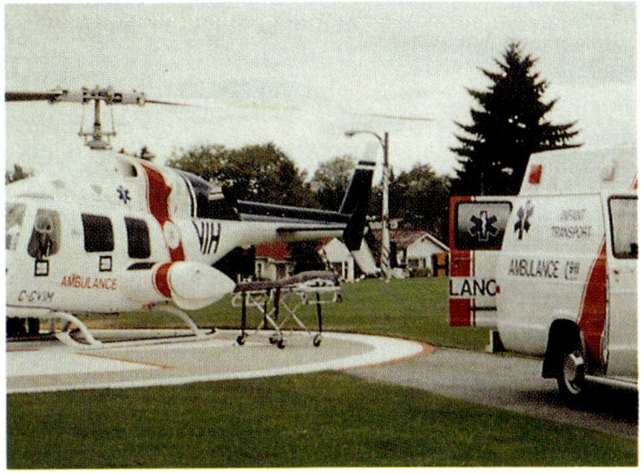

Ambulances that may be either of a standard or specialized type, depending on the circumstances, are used to get patients to the helicopter or from the aircraft to the hospital.

DAVE PARKER

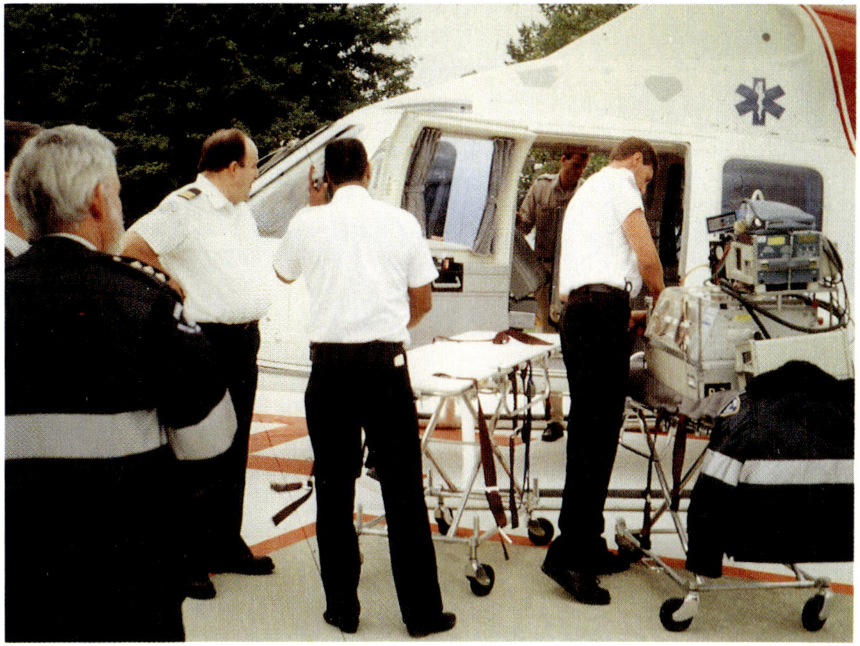

The Infant Transport Team loading an incubator at Vancouver's Children's Hospital on July 22, 1992 for a trip to Campbell River. The paramedic on the right is Andy Goedicke, killed in the crash of a Learjet ambulance early in 1995. Others, left to right, Lynn Klein, Ross Hallaway and Morris Ebank.

DAVE PARKER

The actual development and construction of this device was developed by VIH, working with Sidney-based Automatic Systems Associates Ltd., and with paramedic Randy L'Heureux of the Ambulance Service as a consultant. It was an electrically operated platform that carried the weight of the patient and slid the stretchers into place. The end result is impressive: the helicopter can accommodate a variety of stretchers, incubators of various sizes can be carried and there are two pivoting/sliding attendant seats – in short, the machine now offers all the equipment to be found in ground-based ambulances, and meets the 2-2-2 stipulation.

The 222-UT especially is popular with pilots; the performance is impressive, according to Bill Buhr:

> [The 222] is just a pilot's dream come true – unbelievable! I've flown just about everything and it's got so much reserve power, even single-engined. You can be totally grossed out, and if you happen to lose an engine, which is a very, very infrequent event . . . you can go into a hover with a full load on the remaining engine. There's no other twin-engined aircraft I know that can do that. The [UT's] got 100 hp more a side [than the 222A] – that's 750 hp a side. It's a huge difference.

The Ambulance Service came up with a distinctive paint job and trademark: "Sound of Life." This livery readily identified the helicopters as ambulances and reduced the number of complaints about noise. By 1992, VIH had acquired two additional aircraft in Japan and one in New York, and equipped them with the Easy Loader and all the equipment required by the paramedics. Operations with the 222s began in July of that year.

Crews in this service receive extensive training and have to be highly motivated. The pilots must have instrument ratings and have to be prepared to fly in almost any weather, yet still maintain stringent safety standards. The paramedics, already qualified in advanced life-support roles, must take a further 80 hours of training in the particular problems posed by flying – such problems as changing pressure with altitude and the potential effect on patients.

Paramedic Randy L'Heureux describes an operation to North Pender Island:

> We heard that there was an unconscious [stroke victim]. We headed for the VIH hangar and climbed on board the helicopter. This was a staged response – one where you were sitting ready but hadn't yet had the order to go. The order came quickly and we were on our way to the island. There's a pad there and we were met and transported

> There are four machines but there are 19 pilots. It [the ambulance service] started out to be a logistical nightmare. It's just now [1995] beginning to look after itself – it's running well.
>
> *Bob Reimer*, Chief Pilot

DAVE PARKER

DAVE PARKER

Randy L'Heureux.

to the scene. The patient was unconscious and vomiting profusely, aspirating his vomit. Fortunately for the patient there was a doctor at the scene treating him when we arrived. The patient was prepared and we made a fast trip to hospital providing suction all the way.

The patient survived without permanent impairment. In cases like this, there is little doubt about the efficacy of the helicopter service. Many patients remain eternally grateful to the air ambulances and their crews whose presence has saved their lives. Between them, VIH and the B.C. Ambulance Service have created an economical and efficient air ambulance out of what many had initially considered a less than ideal aircraft for the purpose; and now VIH is marketing the medical conversion kits with the Easy Loader, and there have been sales, along with the reconfigured aircraft in some cases, in the potentially lucrative U.S. market.

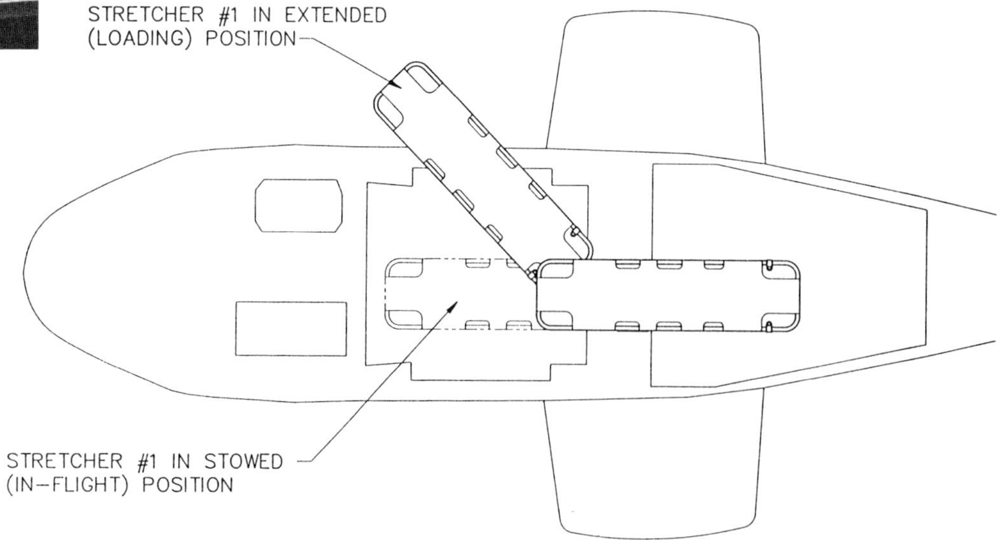

Easy Loader.

VIH COLLECTION

H375
VIH
CCCP

CHAPTER FIVE

Helicopter Logging: Another New Dimension

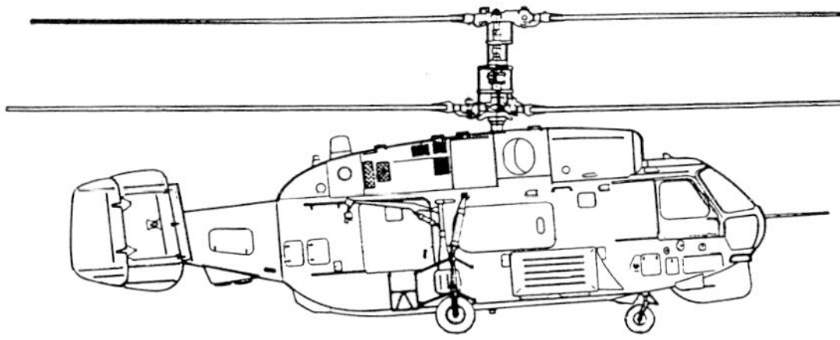

The term heli-logging describes simply the removal of logs from the woods to the nearest dump site after they have been felled; and the first experimental heli-logging operation in British Columbia was carried out more than 20 years ago. Okanagan Lifts Ltd., a subsidiary of Okanagan Helicopters, logged the Richmond Plywood quota on East Redonda Island, 27 miles (44 km) northeast of Campbell River. The 7,500-pound capacity (3 400-kg) Sikorsky S-58T flew an average of 5½ hours a day and 1,400 logs were hauled the two miles (3.2 km) down to the dump site in Pendrell Sound. The three-week experiment, carried out during June 1971, was successful if marginally profitable.

For this reason, it was difficult to justify the use of helicopters in this role in the past. However, during the last two decades, easily accessible timber has dwindled, and environmental concerns about the damage wrought by logging roads – slides and stream degradation, for example – has made logging by helicopter more and more viable in spite of the high costs. Consequently, it seems almost inevitable that Vancouver Island Helicopters would move into this field – inevitable because the president, Frank Norie, has spent his entire career in the logging industry.

Frank describes himself as a pen-pusher before he decided to get into logging; but, in what were still the Depression years, it was not easy to break into the industry – particularly if you lacked any experience in the woods. Eventually, he managed to get a job as a millhand at a small portable mill, with a connected logging operation. "The chain of command there was that if you got a job in the mill, you could work your way up and eventually get a job logging in the bush. So I started as a slab-packer on the green chain, and did work my way up to become a whistle punk in the logging show."

VIH Logging Ltd.

◁ The Kamov KA-32's coaxial rotors leave impressive vapour trails in their wake. Essentially Russian markings were retained during the Kamov's highly successful trials in B.C. Canadian pilots and crews were very favourably impressed.

VIH COLLECTION

By 1942, he and a partner won a contract to cut poles; then came the opportunity to purchase an old but serviceable donkey engine, and a little patch of timber. "So we were away, we got started on our own." A year later, the partner pulled out and Frank was joined by his brother Henry. From then on the brothers never looked back. They became involved in much larger logging ventures and prospered. During this period, they came to appreciate the value of aircraft to get from one work site to another, to seek out promising timber stands and to survey routes for logging roads. In 1957, they purchased a Stinson 108-2, followed by a Seabee amphibian, a Bellanca Cruisemaster and finally a Bell 47 G2 helicopter; and both Frank and Henry became licenced pilots.

They had always done their own engineering: locating timber and marking the boundaries; then the roads had to be laid out and the falling boundaries marked. The Seabee certainly helped, but there were many places where a good deal of travel had to be accomplished on the ground. Frank offers one example from logging country north of Campbell River on the east side of the Island:

> We were going to log off a good sized patch on the side of a mountain by Roberts Lake and, to get into it, we cleaned out an old logging grade, enough for a Jeep to get through. That took us to one arm of Stella Lake. We had a little dinghy there, with an outboard, which took us to the closest spot to that mountain we needed to work on and then we had to walk through the bush – thick second growth which had grown up just about as thick as hair on a dog's back; so we had to more or less cut our way through it the first time. Even then it took a long time. Usually it was about noon before we arrived, where we wanted to do a day's work. After we'd got everything laid out and made our application, the government went in with cruisers, and they did it the easy way. There were some nice, flat, open rocks, right in the middle, and they landed there in a helicopter.

While Henry was reluctant, Frank eventually persuaded him and they bought their first helicopter, a Bell 47 G2, from Alpine Helicopters of Calgary. In the end Bruce Payne taught the whole Norie family to fly: Frank, his son, Ken, Henry and Henry's wife Brenda. The G2, CF-KGS, is still being used by Frank to commute from his home near Duncan to the VIH hangar. Frank's son, Ken, began flying in his teens. He had a private licence on his 16th birthday, and a commercial licence with a helicopter endorsement on his 18th birthday. So there was ample expertise for the company to move into helicopter logging.

Ten hours is a long day, but we don't have a captain and a co-pilot – we change seats every 75 minutes, which helps. The big thing is the rotation. With heli-logging, we do two weeks on and two weeks off and this helps to recharge your battery. Also, of course, it gives you some family life.

Jim Stone, Pilot

We provide the support for heli-logging with our 206s . . . and it's not just for VIH Logging. Canadian Air Crane, for example, that's running three sky cranes [Sikorsky S-64s] and we do their support work. It starts off with us flying in the fallers, perhaps three weeks, perhaps six weeks before the hauling begins. Then, when the Sky Crane turns up, we supply the 206 that flies the chokers and the hill crew to keep things moving.

Charlie Mooney, Operations Manager

I did quite a bit of heli-logging with [Sikorsky] S-58s – the turbine version with derated power, because full power tended to twist the tail off. We worked all round B.C. Eventually I got onto the 61, and that makes you feel you're really moving timber. It'll be even better if we get the Kamov.

Murray Conway, Pilot

When VIH first became interested in heli-logging, it contacted a company operating a Super Puma, one of very few of these helicopters used in logging, in the Campbell River area. The hope was for an arrangement to share the aircraft, which was not at the time being fully utilized. Before this, however, the Puma, while lifting a load of logs off a sidehill, crashed. Both pilots were killed and because the accident resulted from a structural failure, any thought of using a Super Puma in logging was dropped.

VIH, having agreed to move logs for a timber company, had to look for another machine. There was a Sikorsky S-61N, CF-MQF available, but its owner, Bill Thomson, was using it to haul logs from a stand of timber he owned near Prince Rupert. Then one day, Ken Norie, now president of VIH Logging, received a call "out of the blue" from Thomson saying that he couldn't justify hanging onto the helicopter operation and wanted to sell the S-61 and two JetRangers. This was a surprise because Thomson had claimed that he intended logging "forever." Ken describes what happened next.

> Barry [Hewko], my Dad, and I put some serious thought into it and decided that it was more than we could handle financially. . . . It just so happened that Chum [Carley], who owned a Fairchild Hiller 1100, happened to be in the hangar one Saturday morning when Barry and I were here, and Barry said "Chum, you're in logging, how'd you like to buy an S-61 and lease it to us?"

Chum Carley did buy the machine and VIH got Thomson's helicopter company and the two 206s. Leasing the Sikorsky S-61, VIH sent it off to its first job in Sechelt in 1988. This lease arrangement continued for two years, and when Chum Carley was killed in the crash of his FH-1100, VIH purchased the big Sikorsky from the estate.

The log market soon softened and there were problems integrating the big S-61 into the VIH fleet – pay scales, work schedules and differing approaches with the charter operation were difficult to harmonize. By 1991, the Nories and Hewko decided that these problems and the cost of operating the 61 made it unacceptable and they sold the machine. Although they were now out of the heli-logging business, Ken Norie, along with one or two others who had been involved with the heli-logging, were convinced that they could make a go of it with a better market and the increasing need to go after less accessible timber. As well, they would have to work with their own timber stands, rather than act as a flying truck for others.

The Sikorsky S-61N, used by VIH Logging on the Grant Lake logging operation in the spring of 1995. DAVE PARKER

The S-61 undergoes inspections, routine maintenance and refuelling after three hours flying time. All maintenance and support is provided by VIH Logging themselves. Engineer Jim Sawzcuk is shown working between flights.
DAVE PARKER

VIH Logging pilots Mike Long, Jim Stone and Murray Conway. Mike flew the 206 support helicopter, and Jim and Murray the S-61. VIH COLLECTION

Taking off after its maintenance, C-FQNG hoists its 250-foot (72-m) longline and hook. Choker cables are flown to the logging site by a 206. DAVE PARKER

VIH's first S-61N in the parent company's livery. At this stage, VIH Logging was not a separate company. VIH COLLECTION

The Kamov crew: in red cap, Boris Trushin; left to right in cockpit: Bill Ross, Nikolai Kozyrev and Jim Stone. VIH COLLECTION

Around 1989 – to avoid the earlier difficulties, they formed a separate company: VIH Heli Logging. Both Ken Norie, president of the new company, and Bill Ross, the operations manager, went on the hunt for a more suitable helicopter: they found it in the Russian-built Kamov KA-32. Ken describes how they did this.

> We'd heard various stories about these Russian Kamov helicopters that were working in New Guinea; and there had been one of them at the Abbotsford Air Show in 1986. . . . [We felt] there was a potential for this machine to haul logs – It could haul 11,000 pounds [4 900 kg], but there was still a lot of scepticism on our part that a Russian helicopter could work 200 hours a month. To find out what we could, we spoke to the people involved in the New Guinea operation and they said, "Why don't you come over?" So we did.
>
> We spent a week in Papua New Guinea, at Mount Hagen, near the centre of the island and rode around in the thing and Bill flew it [with a Russian check pilot], and talked to the people who were using it and to the Russians who were flying it . . . The only problem that they were having was that the Russians wouldn't allow the westerners to fly it operationally – but they'd had no maintenance problems and the machines were reliable.
>
> After that trip to New Guinea, we realized that what they were doing there wouldn't work for us. We had to have more control so that the guys flying would get more production. We'd already communicated with the Kamov Design Bureau, to start with at a Helicopter Association International convention in 1991. Every year the Russians put on what they called a Russian Symposium which involved speeches by various Russians to show what they were doing with helicopters. After Sergei Mikheyev [Kamov general designer] gave his speech, I stayed behind, waited until everyone else had left the room and then found his translator. I told him who I was and that we were interested in the KA-32 and what we were interested in using it for.

Mikheyev was very interested in doing business, in getting their helicopters out of Russia and on to the world market. At the Russian's request, Ken sent him a letter which, strangely, was never answered. Months passed and Norie and Ross had, in the interim, made their trip to New Guinea. The problem may have been that the people in the KAMOV Design Bureau were very inexperienced in dealing with a situation of this kind. They in fact had never had to do this since previously, under the by now defunct Soviet system, aircraft sales had been carried out by an arm of the government, AVIAEXPORT. The designers were not businessmen.

Ken then went to Russia to try to negotiate a lease, explaining to the Russians what VIH planned to do. Before he went, VIH also had a commitment from Transport Canada that they would be allowed to test the Russian

BOB TURNER

The uncluttered instrument panel of the D1.

When we started flying it [the Kamov], all the other operators thought it was a joke that would be here for a little while then leave, but after the first couple of weeks, rumours started getting out from people in the logging industry about what it was actually doing, how much wood it was moving, and how it had never broken down. People started to get excited, and questions started to be asked as to why it had ever been allowed to come here in the first place.

Ken Norie, President, VIH Logging

BOB TURNER

The panel of the Kamov shows far greater sophistication.

machine. The negotiations were tough; the Russians wanting to be paid more than $2,000 (U.S.) more an hour than VIH Logging could afford based on logging conditions and the market at the time. Negotiations continued over the next month or two, and at last the Russians agreed to send the machine over for a trial operation. As it turned out, they sent the KA-32 over here without a written agreement of any kind – it simply arrived. The agreement was signed later, in 1991. Ken relates the next move:

> Air Show Canada [at Abbotsford] had arranged to have this big Russian transport come over. Kamov managed to get the helicopter loaded on one of those Ilyushins in Moscow. Until the plane arrived, we didn't know if the helicopter would be on board – but sure enough it was there and the maintenance crew was there, too, assembling it. None of them spoke English; they didn't really know who we were, but I pulled out a card with "VIH" on it, and they reacted like we were old friends! We got hold of a translator and they said yes, they'd brought the helicopter over for us and they were getting it ready for flight.
>
> They had a pilot who was going to fly it for the air show, and then what did we want to do with it? I said, "we can't leave it at Abbotsford and so we'd better fly it to Victoria." Their pilot didn't speak English, but through the translator, we managed to get across that Bill [Ross] had flown one a little bit and basically knew how to fly the thing and that we wanted the Russian pilot to fly the thing with Bill in the right seat working the radios and so on. I took Bill over in the Cessna 337 and he hopped in with the Russians and the Russians started it up and flew it back here.

No one in the heli-logging industry believed that it would happen. There was considerable scepticism. Even before it had arrived, VIH had decided that the existing cargo hook arrangement was inadequate and that a western style hook would have to be installed. The dimensions of the hook mount desired was sent back to the Kamov engineers and they built a hook-mounting system out of titanium that adapted to their own mounts in the fuselage. Everything came from Russia with the helicopter, and all the parts, western and Russian, fitted together beautifully. VIH had to install an instrument to indicate the load weights on the hook, and a new release system so that a load could be punched off easily – this took a week.

For the Russians, a busy month of flying was 80 hours, far less than the 150 required in logging to make the operation viable. Ken continues:

> We started flying on about the third or fourth of September, and by the end of September, we'd put on about 200 hours. The Russians, both the ones here doing the maintenance and the others in Moscow, just couldn't believe that this was possible. We had two Russian pilots, and before they left Russia, they were given authority to

> endorse – basically to give permits – to Bill Ross and Jim Stone. Their pilots came over without ever having had a longline on their helicopter and so they just rode the right seat; Bill and Jim did all the flying. [The Kamov] was operating as a foreign-registered aircraft and so [licensing] was out of the DOT's hands; we were working under Russian regulations rather than Canadian.

Originally, this test was to have lasted 90 flying days. VIH Logging decided very quickly that the Kamov was the machine the company wanted, and at that point the company sent a letter to Transport Canada Headquarters in Ottawa requesting type approval (a Canadian airworthiness certificate). The response was that it had never been done before – a Russian helicopter had never been approved in the West. Mikheyev came to Canada, and then he and Ken Norie went to Ottawa to talk to senior officials.

They were told that it would take only about three months. Kamov put in an official request for type approval which resulted in a delegation going to Moscow from Transport Canada. VIH Logging persisted but senior Transport Canada officials refused an extension on the grounds that it was unfair competition. But VIH was logging its own timber, there was no competition; consequently, the extension was granted.

Kamov proved that allegations, by both Canadian and U.S. operators, claiming that the Kamov KA-32 was unsafe were simply untrue. It dropped the campaign for further extensions altogether and concentrated instead on attempting to get type approval. This effort continues.

Meanwhile, logging is continuing, but this time with the S-61Ns, C-FQNG and CF-TIG. In April 1995, operations were under way to the west of Shawnigan Lake off the Renfrew Road. Logs were lifted off the sidehill, hauled by the S-61 to a clearing beside the road where they were loaded onto trucks for the trip to either Pacific Forest Products' Ladysmith dry land log-sorting facility or the Fibremax operation in Victoria.

Sergei Mikheyev.

KAMOV

EPILOGUE

Of the pioneers who started this adventure called helicopter operations, now more accurately described as an industry, some have left us; others are still very active. Sadly, Ted Henson was the first to go in 1957, followed by Carl Agar in 1968. There is regret that we never had the opportunity to interview Carl Agar – though Peter Corley-Smith did meet him in the early 1960s. Later we lost Bob Taylor and Art Johnson. Alf Stringer, however, two years shy of 50 years in the business, is still very much involved as a consultant to VIH; and Barney Bent is enjoying a well-deserved retirement in White Rock. Lynn Stringer, who ran VIH when her first husband, Ted, went down in Tatlayoko Lake, still pursues her many interests, one of which is to teach and then encourage young people to play championship chess.

As for the helicopters, they have progressed from the little one-passenger Bell 47 B3, to machines capable of carrying more than 45 passengers in the cabin, or lifting 30,000 pounds (13 000 kg) on the hook. From crop spraying, their abilities have multiplied until it would require several pages to catalogue all the tasks they can accomplish. The spectrum stretches from collecting fir cones to lifting heavy equipment onto high-rise buildings; from carrying sightseers over the Bugaboo Glacier to ferrying roughnecks to offshore oilrigs. However, their principal advantage lay, and still lies, in reaching isolated areas for resource development; and this inevitably leads to increasing human visitation if not habitation – which, in turn, is usually detrimental to wildlife and its habitat. On the positive side, heavy drilling equipment for oil exploration can be put in place without building damaging roads in sensitive environments; and logs can be retrieved without the roads that cause slide damage, wash-outs and the consequent degradation of soil and fish habitat.

In recent years, helicopters have begun working in less isolated areas. Helijet Airways has run a scheduled service between Vancouver and Victoria, one of the first successful and enduring uses of the helicopter as an airliner. As well, helicopters have begun to play an increasingly important role in tourism. A number of companies offer hiking tours in the wilderness in the summer months. Groups of hikers are put down, often high in the mountains, and sometimes left to walk out to ground vehicles; or picked up again at the end of the day. In the winter, heli-skiing – the helicopter replacing and enhancing the ski-lift – has become big business, with an international clientele. One company that specializes in both these recreations, with a number of lodges both in Alberta and B.C., keeps a record of how many vertical feet heli-skiers have logged. Two have logged ten million feet, but the record is held by Ned Damon, with 18 million vertical feet (6.6 million metres).

Carl Agar, Alf Stringer or Barney Bent could not have forseen all the future adaptations of helicopter use when they started in Penticton, back in 1947, but with them the story began. The helicopter did finally hurdle the last barrier and penetrate the mountains.

On August 21, 1989, Okanagan Helicopters Ltd. became Canadian Helicopters Ltd.

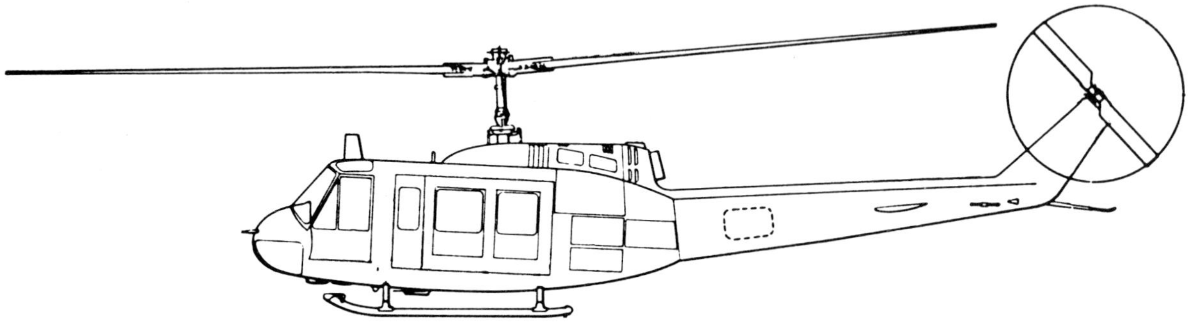

Bell 205A.

INTERVIEWS

Anne Agar
Gordon Askin
Gordon Bates
Barney Bent
Ken Blackwood
Helmut Braditsch
Bill Brookes
Bill Burh
Joan Cannon
Ada Carlson
Murray Conway
Eric Cowden
Trevor Deeley
Ian Duncan
Jock Graham
Barry Hewko
Larry Huff
Art Johnson
John King
Randy L'Heureux
Mike Long
Hugh Lyons
Bill McLeod
Ernie McMinn
Charlie Mooney
Frank Norie
Ken Norie
Bruce Payne
Walter Palubiski
Bob Reimer
Keith Rutledge
Jim Stone
Alf Stringer
Lynn Stringer
Bob Taylor

BIBLIOGRAPHY

BOOKS

Corley-Smith, Peter. *10,000 Hours: A Helicopter Pilot in the North*. Victoria, B.C., Sono Nis Press, 1985 revised 1995.

Corley-Smith, Peter and David N. Parker. *Helicopters: The British Columbia Story*. CANAV Books, Toronto, and The British Columbia Provincial Museum, 1985.

Fay, John. *The Helicopter: History, Piloting, and How it Flies*. Newton Abbot, David and Charles, 1978.

Gablehouse, Charles. *Helicopters and Autogyros: A History of Rotating Wing and V/STOL Aviation*. Philadelphia, Lippincott, 1969.

Gregory, H. Franklin. *The Helicopter: A Pictorial History*. Cranbury, N.J., A. S. Barnes, 1976.

Hubler, Richard G. *Straight Up: The Story of Vertical Flight*. New York, Duell, Sloan and Pierce, 1961.

Sikorsky, Igor I. *The Story of the Winged "S" : An Autobiography*. New York, Dodd, Mead & Co., 1967.

Taylor, Michael J. F., comp. and Taylor, John W. R. ed. *Jane's Pocket Book of Helicopters*. New York, Collier, 1981.

Young, Arthur M. *The Bell Notes: A Journey from Physics to Metaphysics*. New York, Delacourt Press, 1979.

ARTICLES

Edwards, Les. "Okanagan Helicopters." *Air World*, March/April 1967, pp.128-130.

Emerson, G. C. "Topographic Mapping by Helicopter." *Canadian Surveyor*, vol. X, no. 11, January 1952, pp. 3-11.

Graf, Cherry. "Heavy Lift Division Logs." *Rotortales*, vol. 2, no. 4, August 1981, pp. 1-3.

Hanson, Phil. "West Coast Helilogging: Reaping the Harvest with Rotary Wings." *Canadian Aviation*, vol. 52, no. 8, August 1979, pp. 212-24.

Keith, Ronald A. "Okanagan Helicopters: World Wide Operations with 161 Whirly-birds." *Exxon Air World*, vol. 32, no. 2, 1980, pp. 26-29.

Klemin, Alexander. "Sikorskys Helicopter." *Scientific American*, vol. 163, no. 1, October 1940, p. 189.

Slocomb, A. G. "The Helicopter on a Photo-Topographical Survey." *The Canadian Surveyor*, vol. X, no. 1, July 1949

Tomey, Ralph. "Logging by Helicopter in B.C.: An Overview." *Canadian Forest Industries*, vol. 120, no. 3, March 1980, pp. 14-16.

Tower, Courtney. "Too Busy to Fly." *Canadian Aviation*, vol. 30, no. 1, January, 1957. p. 56.

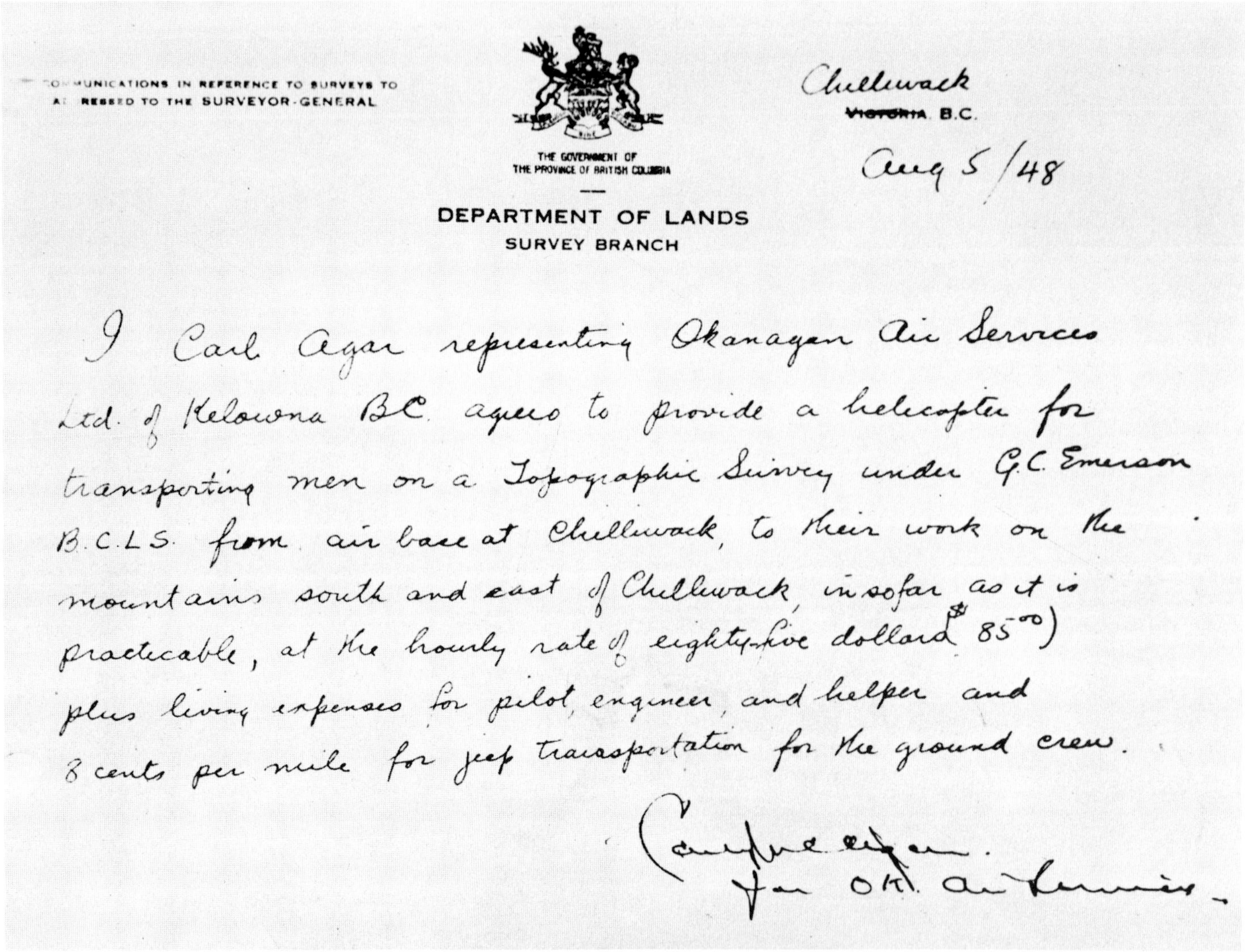

COMMUNICATIONS IN REFERENCE TO SURVEYS TO
ADDRESSED TO THE SURVEYOR-GENERAL

THE GOVERNMENT OF
THE PROVINCE OF BRITISH COLUMBIA

DEPARTMENT OF LANDS
SURVEY BRANCH

Chilliwack
~~VICTORIA~~ B.C.

Aug 5/48

I Carl Agar representing Okanagan Air Services Ltd. of Kelowna B.C. agrees to provide a helicopter for transporting men on a Topographic Survey under G.C. Emerson B.C.L.S. from air base at Chilliwack, to their work on the mountains south and east of Chilliwack, insofar as it is practicable, at the hourly rate of eighty-five dollars ($85.00) plus living expenses for pilot, engineer, and helper and 8 cents per mile for jeep transportation for the ground crew.

Carl Agar.
for OK. Air Services.

Okanagan Air Services first Topographic Survey contract, August 5, 1948.

OKANAGAN COLLECTION

INDEX

A

Abbotsford, B.C., 83
Abbotsford Air Show, 82, 83
Adams, Ron, 41, 49
Aero Copters Inc., 30, 32, 33, **37**, 43
Agar, Carl, **10**, 11, 12, 13, 14, 15, **16**, **18**, **19**, **20**, **21**, 23, 24, 25, 26, 27, 28, 29, 32, 46, 85, 86
Air Transport Board, 13, 30, 48
AIRCRAFT
  Fixed Wing
  Canadair North Star, 33
  Bellanca Cruisemaster, 78
  Cessna 337, 83
  Consolidated Canso, **37**, 39, 44
  de Havilland Tiger Moth, 12
  Ercoupe, 29
  North American Harvard, 33
  Ilyushin, 83
  Learjet, **73**
  Republic Seabee, 78
  Stinson 108-2, 78
  Rotary Wing
  Aerospatiale Super Puma, 79, 20, 26, 29, 32, 41, 45, 48
  Bell 47, 63
  Bell 47B, **16**
  Bell 47 B3, 14, 15, **16**, **19**, 23, 64, 85
  Bell 47 D1, **19**, **21**, **22**, 24, 26
  Bell 47 G, 30, 31, 32, **37**
  Bell 47 G2, 45, 49, 66, **67**, 78
  Bell 47 G3B, 65
  Bell 47 G3B-1, 64, 67, **50**, **52**, **53**
  Bell 47 G3B-2, 64, 66
  Bell 47 G4, 65
  Bell 47 J2, 52
  Bell 204B, **51**, **58**, 65
  Bell 205A, 65
  Bell 205A-1, **55**, **58**, **59**
  Bell 206, **46**, **53**, **55**, **56**, **58**, 60, 62, 64, 65, 67, 78
  Bell 206B, 65, 67
  Bell 206L, **59**, 67
  Bell 206L-1, 65
  Bell 206L-3, **59**
  Bell 212, **68**, 70, 71
  Bell 222, 70, 71
  Bell 222A, 67, 70, 72
  Bell 222-UT, **61**, 67, **71**, 72, **73**, 74
  Fairchild Hiller 1100, 79
  Hiller 12E, 48
  Hiller 360, 26
  Hughes 500C, 65
  Hughes 500D (now McDonnell Douglas), **54**, 65, 67
  Kamov KA-32, **76**, 79, 81, 82, 83, 84
  MBB (Messerschmitt-Blkow-Blohm), 105, 70
  Robinson R22, 22, 49, 55, **66**, **67**
  S.A. Alouette III, 44, 64
  Sikorsky S-51, 26
  Sikorsky S-55, **21**, 28
  Sikorsky S-58, 60, 61, 62
  Sikorsky S-58T, 77, 79
  Sikorsky S-61N, 67, 79, **80**, **81**, 84
  Sikorsky S-64, 78
  Sikorsky S-76, 71
Atkens, Ormonde St. P., 13
ALC Airlift Ltd., 69
Alice Arm, B.C., 47
Alpine Helicopters Ltd., 65, 78
Aluminum Company of Canada, **21**, **22**, 28
American Helicopter Association, 64
Atlin, B.C., 39
Automatic Systems Associates Ltd., 74
Avery, Eugene, 49
AVIAEXPORT, 82

B

B.C. Ambulance Service, **61**, **68**, 69, 71, 74, 79
B.C. Electric Company, 27
B.C. Forest Service, 23, 30, 33, 41, 43, 44, 47, 49, 60, 62
B.C. International Engineering, 28
B.C. Power Commission, 30, 31, 33, 41
B.C. Surveys and Mapping Branch, 25, 31
B.C. Topographic Survey, 17, **19**, 26, 28, 33, 43, 86
B.C. Power Commission, 39
Bates, Gordon, 71
Beebe, Joe, 14
Bejnar, Roman, **73**
Bell company, 13, 14, 15, 27, 29, 32, 64
Bent, Barney, 11, 12, 13, 15, **20**, 23, 85, 86
Board of Transport Commissioners, 31
Boeing, Bill Jr., 30, 31, **36**, 40, 66
Borden, Don, 47, 48
Bowser Lake, B.C., 26
Brady, Carl, 14, **16**
Bridgeport, Connecticut, **21**
British Columbia, 14, 28, 33, 41, 60, 61, 62, 70, 77
British Columbia Forest Act, 62
Brown, Craig, **66**
Buckerfield, Ernie, 13, 15
Bugaboo Creek, B.C., 33
Bugaboo Glacier, B.C., 85
Buhr, Bill, **61**, 74
Burns Lake, B.C., 47
Bute Inlet, B.C., 32
Butler, Gordon, 13

C

Calgary, Alberta, 29, 33, 43, 78
California, U.S.A., 66
Campbell River, B.C., 33, 48, 62, **73**, 77, 78, 79
Canadian Air Crane Ltd., 78
Canadian Aviation, 15, 39
Canadian Forest Service, 62
Canadian Helicopters Ltd., 86
Canadian Pacific Airlines, 13
Canadian Surveyor, 15
Carley, Chum, 79
Cassidy, 62

Central Aircraft, Inc., 14
Chapman, Mack, 33
Cheam Peak, B.C., 26
Chilliwack, B.C., 25, 26, 33
Chilliwack Lake, B.C., 25
Chukachida Lake, B.C., 39
Comox, B.C., **73**
Conway, Murray, 79, **80**
Corley-Smith, P., 31, 47, 48, 85
Courtenay, B.C., 62
Cowden, Eric, 39, 64

D

Dawson Creek, B.C., 33
Deeley, Trevor, 55, 67
Department of National Defence, 33
Department of Transport, 30, 31, 33, 44, **54**, 84
Detroit, **71**
Dewar, Douglas, 13, 14, 15, 26
Dodwell, Noel, 41, 42, **43**
Dominion Experimental Station, 14
Dominion Geodetic Survey, 15, 26
Duncan, B.C., 78
Dunn, Dan, 47

E

East Kootenays, B.C., 23, 33
East Redonda Island, B.C., 77
Ebank, Morris, **73**
Emergency Health Service Commission, 69
Emerson, Gerry, **17**, **19**, 25, 26, 27
England, 70

F

Fairchild F-24 camera, 49
Falconbridge Nickel, 33
Federal Aviation Administration, 13
Federal Department of Mapping and Surveys, 26
Foote, Bill, 48
Forest Industries Flying Tankers, 62
Fornoff, Art, 29
Franklin engine, 14
Fraser River, B.C., 33
Fraser Valley, B.C., 23
Freyney, Father, 12

G

Gagnon, Lou, 12
Gardner Canal, B.C., 28
General Aircraft, 12
Goedicke, Andy, **73**
Golden Hinde, B.C., 33
Goodall, Howie, **37**, 41
Gordon Head, Victoria, B.C., 46
Gott Peak, B.C., 27
Grady, Jim, **50**, 61
Graham, Jock, 28
Grant Lake, B.C., **80**
Grant, Ron, 30, 31
Great Central Mountain, B.C., 33
Great Northern Hotel, 12
Greater Vancouver Water District, 27
Gulf Islands, B.C., 71

H

Hall, Bill, 49
Hallaway, Ross, **73**
Harley-Davidson, 67
Hart Highway, B.C., 33
Hazelton, B.C., **19**
Helicopter Association International, 82
Helijet Airways, 85
Henson, Danny, 40
Henson, Lynn, 29, 30, 31, **35**, **36**, **37**, 39, 40, 41, 45
Henson, Ted, 29, 30, 31, 32, 33, **34**, **36**, **37**, 39, 40, 45, 85, 87
Heslop, Dr. Wilfred, 27, 28
Hewko, Barry, 49, **49**, 66, 67, 70, 71, 79
Holden, Dave, **56**
Homakko River, **34**
Homathko River, B.C., 49, **56**, 31, 32, 33, **37**, 39, 41
Honeyman, Don, 66
Horetsky Creek, B.C., 24
Howroyd, Joe, 31
Huff, Larry, 66, 70, 71

J

J. C. Sproule & Associates, 43
Japan, 74
Jeffries, Ron, 66
JetRanger, 65, 67, 79
Joans Flower Shop, **58**
Johnson Street Bridge, **46**
Johnson, Art, 85
Jones, Harvey, 32, **34**, **36**, 41

K

Kamov Design Bureau, 82
Kemano, B.C., **21**, **22**, 28
Kemano River, B.C., 28
Kenting Aviation Ltd., 15, 26
Kidston, James B., 13
King, John, **60**
Kispiox Valley, B.C., **19**
Kitimat, B.C., **21**, **22**, 28
Kitimat-Kemano project, 28
Klein, Lynn, **73**
Korean War, 32
Kozyrev, Nikolai, **81**

L

L'Heureux, Randy, 74, **75**
Ladner, B.C., 66
Ladysmith, B.C., 84
Lillooet Mountain, B.C., 45
Linhof camera, 49
Long, Mike, **80**
Los Angeles, U.S.A., 29, 62
Lycoming piston engine, **57**, 65
Lyons, Hugh, 49
Lytton, B.C., 27

M

Mackenzie River, NWT, 43
MacDougal, Ian, **66**
Marshall, Dr. James, 14
McGowan, Ray, **54**
McGregor, Fred, 40, 45
McKenzie, geologist 27
McLeod, Bill, **21**, 24, 28
McMinn, Ernie, **19**, 26, 27
McPherson, Glenn, 41
Mikheyev, Sergei, 82, **84**
Monsoon Bucket, **50**, 61
Mooney, Charlie, **62**, 78
Moore, Peter, 45, 48
Moscow, Russia, 83, 84
Mount Douglas, Victoria, B.C., **59**
Mount Hagen, New Guinea, 82
Mount Slesse, B.C., 33
Muraro, John, 62
Mutch, Wally, 14
Mycroft, Brian, **58**

N

Nass Valley, B.C., 28
Nelson, B.C., **50**, 61
New Guinea, 82
New York, 74
Nickel Plate Mine, 12
Norie Bros. Logging Ltd., 66
Norie, Brenda, 78
Norie, Frank, 32, **61**, 66, 67, 77, 78
Norie, Henry, 67, 78

Norie, Ken, **54**, 66, 78, 79, 82, 83, 84
Norman Wells, NWT, 43
Northern Mountain Helicopters Ltd, **50**

O

OHalloran, Des, 31, **37**, 41
Okanagan Air Services, Ltd., 11, 12, 13, 15, 23, 25, 28, 86
Okanagan Helicopters, Ltd., **20**, **21**, 28, **29**, 30, 31, 32, 45, 46, 48, **50**, 60, 61, 62, 77
Okanagan Lifts Ltd., 77
Okanagan Valley, B.C., 12, 14
Ontario, 33, 39, 70
Ostrander, Paul, 23, 27
Ottawa, Ontario, 29, 84

P

Pacific Forest Products, 84
Pacific Great Eastern Railway (later B.C. Rail), 45
Paddy Lake, B.C., 39
Page, Ron, 29
Palisade Lake Dam, B.C., **18**, 27
Pat Bay, Sidney, B.C., 43, **56**, 65, 72, **73**
Payne, J. Bruce, 33, **36**, **37**, 39, 41, 44, 48, **54**, 63, 64, 65, 66, 78
Pendrell Sound, B.C., 77
Penticton, B.C., 11, 12, 13, 14, **16**, 23, 86
Photogrammetric Engineering Journal, 49
Photographic Surveys Ltd., 23
Pitt Meadows, B.C., 69
Poole, Don, 39
Port Hardy, B.C., 48, 66
Port Renfrew, B.C., **37**
Prince George, B.C., 33, 62
Prince Rupert, B.C., 45, 47, 48, **52**, **54**, **58**, 65, 69, 70, 71, 79
Prince Rupert Regional Hospital, **68**
Protheroe, Ted, 32, 33, **34**, **37**, 39, 40, 41

Q

Queen Charlotte Airlines, 27
Queen Charlotte Islands, B.C., **54**, **68**, 69

R

Race Rocks, B.C., **57**
Red Mountain, B.C., **55**
Reimer, Robert, **65**, 66, 74
Roberts Lake, B.C., 78
Ross, Bill, **81**, 82, 83, 84
Ross, Uisdean, **67**
Rossland, B.C., 12
Rotary Winged Aircraft, 26
Royal Canadian Air Force, 11, 20, 26, 29, 32, 41, 45, 48
Royal Jubilee Hospital, Victoria, B.C., 29
Russia, 82, 83

S

Sadler, Don, 47
Sampson, Jim, 23
Sands, Hank, **52**
Sandspit, B.C., **54**
Sawzcuk, Jim, **80**
Scholberg, Don, 32
Seal Cove, Prince Rupert, B.C., 45, 48, **52**, **54**, **58**
Seattle, Washington, 29, 39
Sechelt, B.C., 79
Second World War, 11, 29
Seers, F.P., 15
Shawnigan Lake, B.C., 84
Shelbourne Street, Victoria, B.C., 31
Sidney, B.C., **35**, 48, **55**, 64, 74
Sikorsky, Igor, **10**, 45
Skyways Services, Ltd., 15, **18**, 23, 27
Smithers, B.C., 47
Soledad Canyon, California, 62
Soloy conversion, **57**, 65
Soutar, Al, 15, 23
Southern California Edison Company, 62
Spartan Air Services Ltd., 29, 26, 48
Speed, Frank, **19**
Spence, Howard, 26
Squamish, B.C., 27
Squamish River, B.C., 60
Stella Lake, B.C., 78
Stevenson, Henry, **50**, 61
Stewart, B.C., 25, 26, 47, 48, **52**, **55**, **58**
Stewart, Norman, 25, 26
Stikine River, B.C., 41, 43, 47
Stone, Jim, 79, 80, **81**, 84
Strait of Georgia, B.C., 29
Strathcona Park, B.C., 33
Stringer, Alf, **10**, 11, 12, 13, 14, 15, **16**, 26, 27, 29, 44, 45, 46, 47, 48, 60, 64, 65, 66, 67, 85, 86
Stringer, Lynn, 46, 66, 67, 85
Stuart Lake, B.C., 37
Summerland, B.C., 14

T

Taiwan, 71
Tatlayoko Lake, B.C., **34**, 39, 85
Taylor, Robert P., 32, 33, **36**, 39, 40, 41, 42, 43, 44, 45, 49, 85
Telegraph Creek, B.C., 47
Terrace, B.C., 26, 28
Teslin, Yukon, 15
Tête Jeune, B.C., 33
Thompson, Earl, 33
Thomson, Bill, 79
Thutade Lake, B.C., **50**
Tingstad, Erv, 65, **52**
Tofino, B.C., 27
Toronto, Ontario, 23
Trans-Canada Air Lines, 33
Transport Canada, 82, 84
Trushin, Boris, **81**
Turnbull, Dave, 30
Turner, Bob, 31
Tuttle, A.C, 26

U

University of British Columbia, 27
Utah Mines, Port Hardy, B.C., 48, **17**, **19**
Vancouver, B.C., 11, 12, 13, 15, **21**, 27, 45, 46, 70, 85
Vancouver Island, B.C., 27, 29, 33, 61, **73**
Vancouver Island Helicopters, Ltd., 29, 30, 31, 41, 45, **46**, 47, 49, 64, 65, 66, 67, 69,, 70, 71, 74, 75, 77, 78, 79, 82, 83, 95
Victoria, B.C., 27, 29, 30, 31, 45, **46**, 57, 66, 70, 83, 84, 85
Victoria Flying Club, 30, 33
VIH Logging Ltd., 67, 78, 79, **80**, 82, 83, 84

W

Wahleach Range, B.C., 25
Warren, geologist, 27
Watson Lake, Yukson, 15
Watson, John, 32
Wenatchee, Washington12
White Rock, B.C., 85
Windermere Valley, B.C., 23
Winnipeg, Manitoba, 15, **18**, 23

Y

Yakima, Washington, 12, 13, 14, 15, **16**, **18**